Pharmaceutical Discovery, Development and Manufacturing Forum 2013

Core Programming Area at the 2013 AIChE Annual Meeting: Global Challenges for Engineering a Sustainable Future

San Francisco, California, USA
3 – 8 November 2013

ISBN: 978-1-63439-051-4

Printed from e-media with permission by:

Curran Associates, Inc.
57 Morehouse Lane
Red Hook, NY 12571

Some format issues inherent in the e-media version may also appear in this print version.

Printed by Curran Associates, Inc. (2014)

For permission requests, please contact AIChE
at the address below.

AIChE
3 Park Avenue
New York, NY 10016-5991

Phone: (203) 702-7660
Fax: (203) 775-5177

www.aiche.org

Additional copies of this publication are available from:

Curran Associates, Inc.
57 Morehouse Lane
Red Hook, NY 12571 USA
Phone: 845-758-0400
Fax: 845-758-2634
Email: curran@proceedings.com
Web: www.proceedings.com

TABLE OF CONTENTS

(2a) Introduction

Clarke-Pringle, T. L., DuPont Engineering Research and Technology
Originally presented on: 11/3/2013 13:15:00 - 13:40:00

An introduction to the sessions celebrating the 70th Birthday of John MacGregor

It is with great pleasure that we kick off our special sessions with an introduction of John MacGregor. In our introduction, we will spend some time providing a historical overview of John's key areas of active research: from his early days in polymerization and process control, to his world class developments in multivariate statistics. Many of these areas will be presented in detail by our invited speakers. In the two sessions we will hear from John's students and also from his industrial collaborators. Toghether we honor the monumental scientific legacy that has expanded the theory but also made an impact at the plant floor.

But more importantly, we would like to give our audience a more personal introduction of John MacGregor: a compassionate and kind supervisor, and a person who is successful in life, by all measures. It is with pride and good memories that we call ourselves former JFM students, and we look forward to honoring him in this session.

(2b) A Comprehensive Mathematical Model of Vinylidene Fluoride Emulsion Polymerization in a Semi-Batch Reactor

Kiparissides, C., The Petroleum Institute
Pladis, P., CERTH/CPERI
Alexopoulos, A., CERTH/CPERI
Originally presented on: 11/3/2013 13:40:00 - 14:05:00

Polyvinylidene fluoride (PVDF), the second most important fluoropolymer, exhibits a unique combination of desired end-use properties including excellent chemical resistance, high thermal stability, and good mechanical as well as piezoelectric properties. Due to these unique characteristics, PVDF is widely used in wire and cable insulation, as sheet and melt-cast films in electronics, in fittings, valves and pumps, etc. PVDF is produced commercially by emulsion and suspension polymerization processes.

In the present paper a comprehensive mathematical model of the vinylidene fluoride (VDF) emulsion polymerization process for a semi-batch reactor is developed. The model can predict the molecular (e.g., molecular weight distribution, etc.) and morphological (e.g., particle size distribution) properties of the particulate PVDF in agreement with experimental measurements. The mathematical model includes a comprehensive kinetic mechanism, a gel-effect model based on the generalized free-volume theory, thermodynamic equilibrium models for calculating monomer partitioning in the different phases present as well as the reactor pressure, a physical model for calculating the crystalline and amorphous polymer mass fractions, and a full set of differential equations describing the conservation of various molecular species and energy in the reactor. The method of moments is employed for the calculation of the average molecular weight properties. In addition, a population balance model is included to describe the dynamic evolution of the particle size distribution in the polymerization reactor.

The predicting capabilities of the model are demonstrated by a direct comparison of model predictions with experimental data on the monomer feed rate, monomer conversion, mean particle size and particle size distribution, as well as the molecular weight distribution of PVDF obtained from a pilot-scale polymerization reactor. Detailed experimental measurements on the polymerization rate, monomer feed rate, operating pressure, molecular weight distribution and particle size distribution of PVDF (measured in a 30 lt semi-batch reactor) were provided by an industry. In the experiments, the reactor pressure was kept constant throughout the polymerization via the control of the monomer feed rate into the reactor. The initiator and chain transfer agent were added in the beginning of the polymerization as well as at various discrete times during the polymerization. Simulation results show a very good agreement between model predictions and experimental measurements on the total amount of VDF fed to the reactor as well as with measured particle size and the molecular weight distributions.

(2c) A Multidimensional Spectroscopy Approach for Protein Characterization

Garcia-Rubio, L. H.,

Originally presented on: 11/3/2013 14:05:00 - 14:30:00

Aggregation and oligomerization processes lead to structural changes in proteins that have high relevance in the biomedical and biotechnology fields. Understanding of the mechanisms through which proteins unfold and aggregate is critical for pharmaceutical applications, specifically for the understanding of the relationship between protein structure and drug effectiveness. To date, the information required for this understanding comes from diverse, albeit well established, methods such as crystallography, light scattering (both static and dynamic), circular dichroism, NMR, chromatography and other optical methods including absorption and fluorescence spectroscopy. Because the information from these methods cannot be interpreted concurrently, only partial monitoring of experimental and process variables can be accomplished in real-time for both dynamic experimentation and process monitoring. In this study, a multidimensional spectroscopy platform for the simultaneous measurement and interpretation of multiwavelength angular scattering, orescence, and transmission measurements is demonstrated. This platform s capable of real-time continuous measurements of proteins in solution. The associated integrated interpretation model translates the multidimensional spectral data into useful physical and chemical information such as oligomeric and aggregate types and size distributions, chromophoric amino acid composition, and the response of fluorophores, such as tyrosine and tryptophan, to their local environment. The quantity and quality of the information accessible through multidimensional spectroscopy measurements is demonstrated using NIST traceable particle size standards, and the thermal and chemical denaturation of proteins as case studies. The results indicate that multidimensional spectroscopy is a powerful and exciting approach for accurate, rapid, identification and characterization of proteins and protein related processes.

(2d) Could Polymers Detect Volatile Organic Compounds (VOCs)?

Penlidis, A., University of Waterloo, Institute for Polymer Research

Originally presented on: 11/3/2013 14:30:00 - 14:55:00

With indoor air quality becoming a greater concern, it is important to be able to determine the concentration of indoor pollutants. Formaldehyde is one of the main gases contributing to poor indoor air quality since it is so widely used in the manufacturing of goods. Even at very low (ppb) concentrations, formaldehyde can cause respiratory problems and a general feeling of unwellness. The World Health Organization states that the concentration of formaldehyde over a 30 minute period should be less than 0.08 ppm. Polyaniline (PANI) doped with nickel oxide and/or aluminum oxide was tested as a potential sensing material for formaldehyde using a test system that included a specialized gas chromatograph (GC).

Using formaldehyde as an example, prescriptions are suggested to determine which polymeric sensing materials would be best for any target analyte. These prescriptions are based on the chemical nature of both the sensing material and analyte (in this case formaldehyde) as well as the type of sensor being designed. Immediate extensions of these prescriptions would include ethanol.

(2e) Development and Comparison of Approximate Maximum Likelihood Methods for Parameter Estimation in Continuous Dynamic Models With Stochastic Disturbances

McAuley, K. B., Queen's University
Karimi, H., Queen's University
Originally presented on: 11/3/2013 14:55:00 - 15:20:00

The aim of this work is estimating model parameters in simplified fundamental models intended for on-line process monitoring and state estimation. Stochastic error terms are often included in these differential equation models to account for process disturbances, time-varying parameters and model mismatch. Three maximum-likelihood-based parameter estimation techniques have been developed to estimate the model parameters and the intensity of the stochastic disturbances. These methods, which are designed to be less computationally intensive than Monte Carlo methods, rely on B-spline approximations for the state trajectories. Three different objective functions for parameter estimation have been developed using different approximations for likelihood functions. The first objective function is developed by approximating the expected value of the likelihood for the states and measurements, given the model parameters and disturbance intensities, using the mode of the corresponding probability distribution, assuming that measurement noise variances are known. The second method uses a Laplace approximation for the likelihood function of the measurements given the model parameters, disturbance intensity and noise variances. The third uses a more accurate fully Laplace approximation. These latter methods are more powerful than the first because they can be used when measurement noise variances are unknown. The three techniques, which result in relatively simple objective functions for parameter estimation, are compared with existing approximate maximum likelihood methods using a simple stochastic CSTR example.

(2f) Comparison of Statistical & Mechanistic Models for Batch Process Monitoring

Nomikos, P.,

Originally presented on: 11/3/2013 15:20:00 - 15:45:00

On-line process monitoring is paramount in every industry. It protects against a variety of problems such as to minimize product quality variability, for real time release of the product to the customers, for maintenance of equipment and instruments, and sometimes even for process safety.

Two advanced methods for real time process monitoring will be presented and compared. Both methods use and exploit the information in the on-line measurements that are available around the process. The example that will be used is an industrial Batch process (Nylon-6,6), but the same methods can be applied to continuous processes as well.

The first method is based on Mechanistic models. Mechanistic models are the classical chemical engineering models that incorporate mass and energy balances, kinetics, thermodynamics, etc. These models in the form of Kalman Filters, can be used in real time with the clever use of few process measurements (temperature, pressure, etc) to predict certain states such as product quality,.

The second method is based on multivariate statistical techniques (Multi-way Principal Component Analysis, and Partial Least Squares) that analyze the information contained in the variety of on-line measurements available around a process, and compares this against the normal process behavior that has been identified from a historical database of past successful periods. This result in an efficient real time monitoring scheme.

Both methods will be compared and contrasted, highlighting their advantages and disadvantages.

(26a) Assessing the Relative Merits of Pharma Reactions in Flask Vs Flow

Blackmond, D. G., The Scripps Research Institute

Originally presented on: 11/3/2013 15:30:00 - 15:55:00

The future for the combination of microflow reactors and organic synthetic chemistry for phamaceutical applications is bright, and many novel and creative applications for these reactor systems will find their place in academic and industrial laboratories. However, the decision of whether to run a complex liquid or multiphase organic reaction in a flask or in a microflow reactor must be made on a case-by-case basis. This lecture advocates placing highest priority on a holistic picture including careful study of the process thermochemistry and relevant transport phenomena as well as the reaction chemistry and production requirements in order to make the decision wisely. Discussion points are offered specifically for the case of one-phase, homogeneous catalytic reactions. Mixing considerations are addressed as well as the role of heat transfer, introducing simple calculations that help to assess these issues both at the laboratory and at the plant scale. We argue the case for applying reaction progress monitoring in batch as the most efficient and information-rich means of obtaining a comprehensive kinetic analysis that may then be used to help choose the optimal reactor type.

(26b) Process Intensification of Gas-Liquid-Solid Reactions in the Production of Fine Chemicals With Milli Packed Bed Reactors

Langsch, R., Technische Universität Dresden

Mueller, A.,

Haase, S., Technische Universität Dresden

Lange, R., Dresden University of Technology

Originally presented on: 11/3/2013 15:55:00 - 16:14:00

Process intensification of gas-liquid-solid reactions in the production fine chemicals with milli packed bed reactors

-

1. Introduction

Flavors and fragrances are usually produced by selective liquid phase hydrogenations in batch stirred-tank reactors with suspended catalyst powder. Disadvantages of these processes are an elaborate catalyst separation, a varying product quality in each batch and an inefficient heat transfer so that the educt must be diluted. The transformation of the discontinuous processes into continuous ones using reactors with fixed catalysts and a high heat transfer capability could lead to a more efficient and cheaper production, but requires novel miniaturized reactor systems [1].

Milli packed bed reactors offer high mass and heat transfer rates and seem to be a promising reactor design for the production of fine chemicals by gas-liquid-solid reactions. The reactors consist of parallel flow channels with inner diameters of a few millimeters which are filled with conventional catalyst pellets. In dependence on the channel-to-particle-diameter ratio, dissimilar bed geometries are obtained leading to different hydrodynamics and mass transfer rates [2].

2. Main Objective

The aim of this work was the experimental investigation of overall (gas-liquid-solid) mass transfer coefficients of hydrogen for milli packed bed reactors with different bed geometries for different flow directions. Based on this investigation, the reactor configuration with the highest mass transfer coefficients was applied to study the selective hydrogenation of cinnamaldehyde on Pd/Al_2O_3 catalysts in order to evaluate the applicability of milli packed bed reactors as production unit for fine chemicals.

3. Experimental

Four milli packed bed reactors with different bed geometries were created by filling catalyst spheres with diameters of 0.8 mm and 1.6 mm, respectively, in round channels with different inner diameters (d_{CH}: 1.0; 1.4; 2.0 mm). Schematic illustrations of the packing structures are shown in Figure 1.

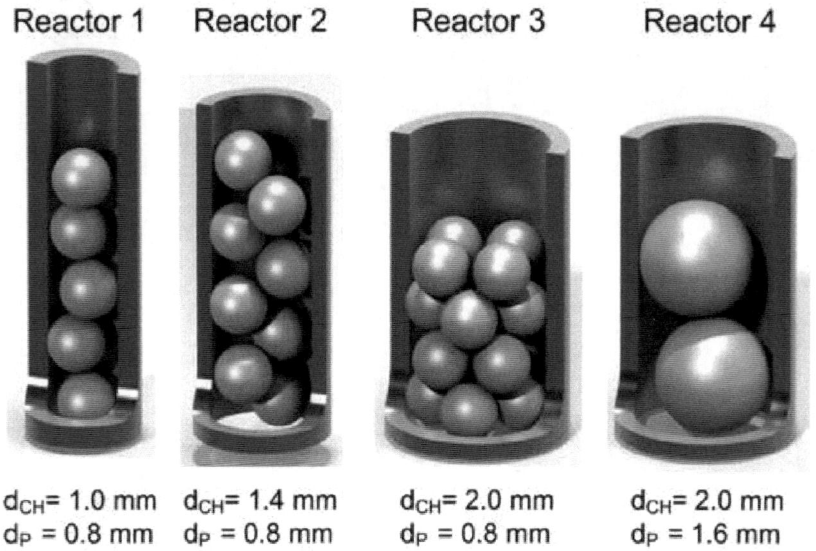

Figure 1 Schematic illustration of the investigated milli packed bed reactors.

These reactors were experimentally investigated with regard to flow patterns and overall mass transfer coefficients of hydrogen under reacting conditions (liquid phase hydrogenation of α-methylstyrene; T: 343-423 K; p: 0.5-1.1 MPa) for different flow directions (downflow, upflow and horizontal flow). The mass transfer studies were performed for a wide range of superficial gas (0.009-1.18 $m \cdot s^{-1}$) and liquid (0.005-0.10 $m \cdot s^{-1}$) velocities.

The selective hydrogenation of cinnamaldehyde was performed in milli packed bed reactors with a channel diameter of 1.0 mm and different reactors with lengths between 0.25 m and 0.55 m in vertical downflow. The temperature was set to 353 K and the pressure was varied between 1.1 and 3.0 MPa.

4. Results

It was found that mass transfer increased with raised fluid velocities in all reactors. For the same catalyst particles (d_P= 0.8 mm), Reactor 1 offered the highest volumetric mass transfer coefficients $(ka)_{OV}$ up to 15 s^{-1} compared to $(ka)_{OV}$= 4.9 s^{-1} for Reactor 2 and $(ka)_{OV}$= 5.0 s^{-1} for Reactor 3 at constant superficial fluid velocities (Figure 2).

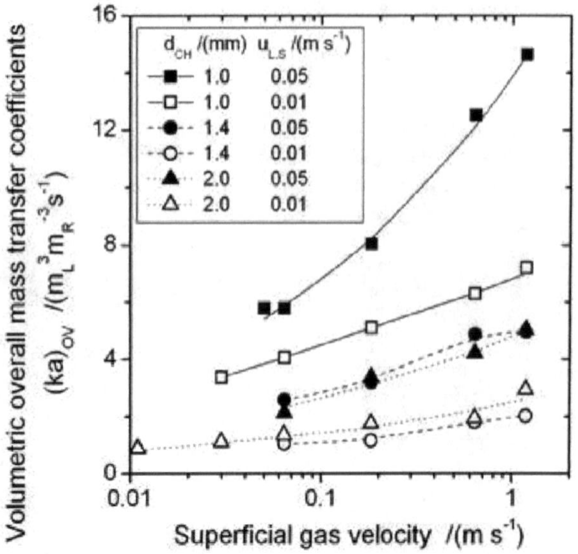

Figure 2 Influence of the bed geometry on volumetric overall mass transfer coefficients for different superficial fluid velocities.

The packing of Reactor 3 and Reactor 4 showed similar volumetric mass transfer coefficients $(ka)_{OV}$, although the volumetric surface area of Reactor 3 (a_{OV}= 3012 m^{-1}) is almost twice as large as Reactor 4 (a_{OV}= 1717 m^{-1}). It can be concluded that the slug flow regime, which is present in Reactor 4 and not in Reactor 3, offers higher mass transfer coefficients k_{OV} compared to the film flow and bubble flow regimes, which are dominate in Reactor 3.

Reactor 2 was used to study the influence of flow direction on overall mass transfer coefficients under the same operation conditions. The experiments showed that downflow offered the highest and horizontal flow mode the lowest mass transfer coefficients for the whole range of superficial fluid velocities. A variation of the flow direction in Reactor 1 gave no difference with respect to the mass transfer coefficients. Such a behavior can be explained by the strong impact of surface tension forces in small channels which make the hydrodynamics independent on the flow regime.

Based on the results of the mass transfer studies, Reactor 1 was applied to study the selective hydrogenation of cinnamaldehyde (CA) to hydrocinnamaldehyde (H-CA) in downflow operation mode. Figure 3 illustrates the conversion of CA and the selectivity of
H-CA related to CA in dependence on the superficial gas velocity for different liquid velocities. It can be seen that conversion of CA increased with raised superficial gas velocity for gas velocities below 0.2 m s^{-1} and decreases slowly for higher velocity. This observation suggests that the external mass transfer is limited for hydrogen at low gas velocities and for CA at high gas velocities. An increased liquid velocity led to smaller conversion due to an

incline of the residence time. The selectivity of H-CA increased with raised gas and decreased liquid velocity.

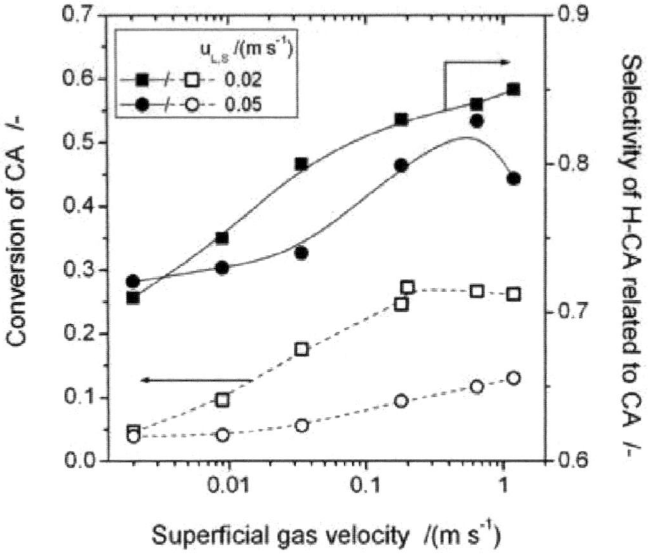

Figure 3 Influence of superficial gas velocity on the conversion of CA (open symbols) and the selectivity of H-CA (closed symbols) for two different liquid velocities (p= 1.1 MPa, T= 353 K, L_R= 0.25 m, 5 % w/w CA in toluene).

Many further reaction studies with different feed concentrations of CA (5 % w/w < w_{CA} < 20 % w/w), hydrogen pressures (1.1 MPa < p < 3.0 MPa) and reactor lengths (0.25 m < L_R < 0.55 m) were performed in order to evaluate the applicability of Reactor 1 for fine chemical syntheses. The results showed that for a feed concentration of 20 % w/w, full conversion of CA can be achieved in a reactor with a length of 1.5 m (p= 3.0 MPa, T= 353 K, d_{CH}= 1.0 mm, $u_{G,S}$= 0.5 m s^{-1}, $u_{L,S}$= 0.1 m s^{-1}). Considering a scale-up of this reactor in order to produce 1 ton H-CA per year, only 3 parallel channels with a diameter of 1.0 mm are required.

5. Summary/ Conclusions

In this work, milli packed bed reactors with different bed geometries operated in 3 flow directions were analyzed with respect to overall mass transfer coefficients under reacting conditions. The selective hydrogenation of cinnamaldehyde was studied in the reactor with the highest mass transfer rates. It can be concluded that milli packed bed reactors can be easily realized due their simple geometry and already a small number of channels can produce several tons per year which make them suitable for application in fine chemical industry.

6. References

1. E. H. Stitt, *Chem. Eng. J.* **2002**, *90 (1-2)*, 47-60.

2. L. E. Kallinikos et al., *Chem. Eng. Process.* **2010**, *49 (10)*, 1025-1030.

(26c) Scale-Up of Oxidation of Alcohols and Aldehyde Utilizing Bleach and Phase Transfer Catalysis (PTC)

Zhang, Y., Iowa State University
Born, S.,
Jensen, K. F., Massachusetts Institute of Technology
Originally presented on: 11/3/2013 16:14:00 - 16:33:00

Using bleach to oxidize alcohols offers several benefits over traditional oxidants: low material cost, mild reaction conditions and no metallic waste. However, the overall reaction rates of such phase transfer catalysis (PTC) reactions are usually limited by mass transfer. Microreactors are known to provide excellent mass transfer rate, but scaling up to production level will need hundreds of microreactors in parallel which can be very expensive. The Low Flow Reactor (LFR) and milli-scale Advanced Flow Reactor (AFR) manufactured by Corning Inc. have larger channel dimensions but still provide comparable mass transport rates. In this work, we compared the observed reaction rate of PTC among spiral design microreactor (0.22 ml), LFR (0.45 ml) and AFR (8.7 ml), and similar conversion was obtained at the same residence time. Through the optimization of the reaction condition and the flow regime, previously mass transfer limited reactions completed within three minutes. Scaling up from microreactor to intermediate devices LFR and AFR without sacrificing mass transport behavior significantly increases production rate and constitute useful flow reactor scaling approaches.

(26d) Use of Modeling and Mechanistic Thinking to Aid Development of a Pd Catalyzed Amination Reaction; Understanding the Mechanism, Oxygen Sensitivity, and Impurity Formation

Merritt, J., Eli Lilly and Co.
Fennell, J., Eli Lilly
Pietz, M., Eli Lilly and Co.
Kallman, N., Eli Lilly
Buser, J. Y., Eli Lilly and Company
Originally presented on: 11/3/2013 16:33:00 - 16:52:00

A Pd catalyzed amination reaction was being developed for use in the synthesis of an API. Reaction rate variability and over-reaction impurities had been observed during development. A combined modeling and experimental approach was undertaken to better understand the chemistry to mitigate impurities and reaction rate variability due to oxygen sensitivity and scale-up effects. Due to the rather complicated nature of the catalytic cycle, a modeling approach was very useful to help disentangle multiple competing effects. In addition to reaction mechanism kinetic elucidation, unit operation modeling will also be discussed to illustrate how it impacts execution of a control strategy for commercial manufacture.

(26e) Robust Hydrogenation Catalyst Selection for Manufacturing of a Pharmaceutical Intermediate: A Kinetic Based Approach

Skliar, D., Bristol-Myers Squibb Co.
Cuniere, N., Bristol-Myers Squibb Company
Brueggemeier, S., Bristol-Myers Squibb
Eastgate, M., Bristol-Myers Squibb
Tummala, S., Bristol-Myers Squibb
Originally presented on: 11/3/2013 16:52:00 - 17:11:00

Selectivity of heterogeneous catalytic hydrogenation reactions often plays an important role in controlling critical quality attributes of pharmaceutical intermediates or API. With the increasing pace of pharmaceutical process development and the greater focus on Quality by Design, there is a growing need to define reaction conditions within a shorter timeframe. Selection of reaction class variables (e.g. catalyst, solvent, additive) can have a profound effect on continuous variable ranges (e.g. temperature, pressure, concentration) and the overall size of the manufacturing process design space. This work will discuss hydrogenation case study where a kinetic based approach to catalyst selection has been applied in the initial phases of process development.

(26f) Development of a Biphasic Organic Solvent System to Broaden the Substrate Range of the Phenylalanine Amine Dehydrogenase

Au, S., University of Kentucky
Bommarius, B., Georgia Institute of Technology
Bommarius, A. S., Georgia Institute of Technology
Originally presented on: 11/3/2013 17:11:00 - 17:30:00

Title: Development of a Biphasic Organic Solvent System to Broaden the Substrate Range of the Phenylalanine Amine Dehydrogenase

Authors: Samantha K. Au, Bettina Bommarius, Andreas Bommarius

The novel enzyme amine dehydrogenase (AmDH) derived through introduction of mutations to the phenylalanine dehydrogenase gene and thus termed F-AmDH, catalyzes the reduction of prochiral ketones to chiral amines [1]. Many desired substrates of F-AmDH show little to no solubility in aqueous media. Organic solvents can be incorporated into the reaction system to improve solubility of the hydrophobic substrates. Dehydrogenases have been scarcely employed in organic solvents because of limited stability of the enzyme in organic media and low solubility of the cofactor. The introduction of water-soluble organic solvents was unsuccessful because of enzyme deactivation in the presence of organic solvents, which led us to transition into biphasic organic solvent reaction systems. In biphasic media, the enzyme and hydrophilic cofactors are envisioned to stay in the aqueous phase while the hydrophobic substrate partitions between the phases but mostly remain in the organic phase. The advantages include a larger amount of total substrate present in the system and absence or reduced substrate and product inhibition.

Hydrophobic substrates were investigated in a variety of organic solvents. We found that for a reaction to occur, a significant amount of the hydrophobic substrate must be present in the aqueous phase. Thus, the best results are obtained with heptane as a second phase in the biphasic reaction system.

In our biphasic reaction system coupled with cofactor regeneration enzyme formate dehydrogenase, the double variant F-AmDH from *Bacillus badius* PheDH creates the product amine of several hydrophobic substrates that were previously unattainable due to low solubility of the ketone substrate in aqueous buffers.Â An example is adamantylethyl-1-amine, obtained from adamantylmethylketone, which reacts with significant rate in biphasic but not at all in monophasic systems.

[1] Michael J. Abrahamson, John W. Wong, Andreas S. Bommarius. *The Evolution of an Amine Dehydrogenase Biocatalyst for the Asymmetric Production of Chiral Amines*. Adv. Synth. Catal., 2013. **355**: p. 1-8.

[2] Michael J. Abrahamson, Eduardo Vazquez-Figueroa, Nicholas B. Woodall, Jeffrey C. Moore, Andreas S. Bommarius. *Development of an Amine Dehydrogenase for Synthesis of Chiral Amines.* Angewante Chemie Int. Ed., 2012. 51: p. 3696-3972.

[3] MÃ©lanie Hall and Andreas S. Bommarius. *Enantioenriched compounds via enzyme-catalyzed redox reactions.* Chem. Rev. 2011, 111, 4088-4110

(26g) Experimental Study and Model of Reaction Kinetics of Synthesis of Methacrylic Anhydride

Ye, C., Fuzhou University

Huang, Z., Fuzhou University

Qiu, T., Fuzhou University

Wang, H., Fuzhou University

Originally presented on: 11/3/2013 17:30:00 - 17:49:00

Methacrylic anhydride is widely used as the esterification agent. In this paper, the reaction kinetics of synthesis of methacrylic anhydride were investigated in a batch stirred-tank reactor at atmospheric pressure by using methacrylic acid and acetic anhydride as raw materials. The reactions were carried out under several temperatures between 343.15 and 363.15K and various starting reactant compositions. Both the influence of temperature and initial molar ratio of Methacrylic anhydride to acetic anhydride on reaction kinetics were studied in a range that is interesting for process design. Base on the characteristics of the reaction, a typical consecutive reaction mechanism was deduced. Homogeneous reversible kinetic model of second-order was described, and the reaction rate constants and activation energy were obtained according to the experimental reaction rate curves. The kinetic equations for the preparation of methacrylic anhydride from methacrylic acid and acetic anhydride are as follows:

$$-\frac{dC_{AC2O}}{dt} = 1.71\times10^{10}\exp\left(\frac{-82.80}{RT}\right)C_{MAA}C_{AC2O} - 8.37\times10^{10}\exp\left(\frac{-88.04}{RT}\right)C_{MAAOAc}C_{HAc}$$

$$\frac{dC_{MAA2O}}{dt} = 5.81\times10^{8}\exp\left(\frac{-78.78}{RT}\right)C_{MAAOAc}C_{MAA} - 2.00\times10^{8}\exp\left(\frac{-69.16}{RT}\right)C_{MAA2O}C_{HAc}$$

According to these equations, the first reaction is an exothermic reaction and the second reaction is an endothermic reaction. The reaction kinetic model was in good agreement with the experimental data. The biggest relative error was nearly 1.5% in the subsequent reaction process.

(14a) Experiences and Lessons Learned From Collaborating With Jfm

Bourg, F.,

Originally presented on: 11/3/2013 16:00:00 - 16:25:00

We gather together to recognize Dr. MacGregor's contributions to our chosen field of endeavor as Process Systems Engineers. Industrial application of Advanced Process Control Methodologies based on Multivariate Statistical methodologies has advanced substantially as a result of the career achievements of Dr. John MacGregor. It is observed that changes of this order are a result of the "character" of the individuals at the forefront of the change, the "tip of the spear". The character qualities that are instrumental in driving changes of the order that have resulted from Dr. MacGregor's work are actually quite simple in concept, but very difficult in practice. I refer to the character qualities as the 3D's of change. Desire, the personal realization that change is required, the "want" to. Dedication, the personal commitment to explore, learn, and develop the tools required to convince others that change is "good", the "will" to do. Determination, the personal commitment to preserver in the face of obstacles and all those who would, in the name of "stability", discourage pursuit of the pathway that is so clear in the mind of the Change Agent, the "commitment" to do. We are gathered here in this venue to celebrate the impact of Dr. John MacGregor in creating and sustaining an environment that not only defined the change he envisioned from an Academic perspective, but also creating the necessary collaborative industrial relationships that have led to realization of changes in practice that we have only just begun to see. Dr. MacGregor's legacy is evidenced, not only in those that he has taught and trained, but also in those that have had the privilege to experience the richness of collaboration in implementation of his visionary concepts and methods into robust industrial applications. As Engineers and Scientists, we realize that our achievements are not ours alone, but are, in point of fact, a result of all those who have gone before us and shone a light into the darkness.

(14b) 20 Years of Applying John Macgregor's Ideas

Champagne, M., Eli Lilly

Originally presented on: 11/3/2013 16:25:00 - 16:50:00

I had the privilege of knowing and working with John over the last 20 years. In honor of John MacGregor's 70th birthday, I will be discussing how his ideas and my interactions with John and his graduate students have influenced me in applying multivariate data analysis (MVA) in the pulp and paper industry, in the steel industry and now in the pharmaceutical industry. I will discuss how multivariate data analysis is applied in continuous manufacturing process in pulp and paper, and in pharmaceutical manufacturing process. I will also discuss Eli Lilly's collaboration with ProSensus personnel in testing John's theories of optimizing existing processes using historical data, and in testing his theories on transferring a product on a different production platform. Finally I will comment about John's thoughts of how the pharmaceutical industry should be developing a better Design of Experiment strategy in the product design space.

(14c) Multivariate Image Analysis and Regression (MIA/MIR) in the Process Industries

Duchesne, C., University of Laval
Liu, J. J., Pukyong National University
Originally presented on: 11/3/2013 16:50:00 - 17:15:00

Multivariate Image Analysis (MIA) was first introduced by Esbensen and Geladi in the mid-late 80's [1] when they proposed to apply Principal Component Analysis (PCA) on multi-channel images. Most of the early MIA research works were made in the fields of remote sensing, analytical chemistry, and medical imaging where multivariate images were available. The history of MIA in the process industries is more recent, and its introduction in the chemical engineering field was pioneered by John MacGregor and his group. It began almost a decade later when Bharati and MacGregor [2] advocated the use of MIA for on-line process monitoring by applying the approach to time series of images. That is using the imaging system as an on-line PAT type sensor for real-time process monitoring (i.e. Multivariate Statistical Process Control). Today, important image-based problems are solved using multivariate approaches such as: 1) monitoring products for defects, 2) predicting overall product quality, 3) feedback control of product quality, 4) monitoring and control of process conditions. Key contributions from the MacGregor's group that helped solving these important process problems include the development of Multivariate Image Regression (MIR) methods for predicting any product property or process condition of interest using spectral information [3], textural information [4] or both, and the use of the image-based models for process control and optimization [5]. These papers led to many other developments and applications in the process engineering field which were recently reviewed in [6]. The objective of this presentation is to provide an overview of the history, methods and applications of MIA that have impacted the process industries. How John MacGregor has inspired further developments in the field, and particularly the work of the two co-authors, will be emphasized.

[1] K. Esbensen, P. Geladi, "Strategy of Multivariate Image Analysis (MIA)", Chemometrics and Intelligent Laboratory Systems, 7, 67-86, 1989.

[2] M.H. Bharati, J.F. MacGregor, "Multivariate image analysis for process monitoring and control", Industrial & Engineering Chemistry Research, 37, 4715-4724, 1998.

[3] H. Yu, J.F. MacGregor, Multivariate image analysis and regression for prediction of coating content and distribution in the production of snack foods, Chemometrics and Intelligent Laboratory Systems 67 (2003) 125–144.

[4] M.H. Bharati, J.J. Liu, J.F. MacGregor, Image texture analysis: methods and comparisons, Chemometrics and Intelligent Laboratory Systems 72 (2004) 57–71.

[5] J.J. Liu, J.F. MacGregor, Modeling and optimization of product appearance: application to injection-molded plastic models, Industrial and Engineering Chemistry Research 44 (2005) 4687–4696.

[6] C. Duchesne, J.J. Liu, J.F. MacGregor, "Multivariate image analysis in the process industries: A review", Chemometrics and Intelligent Laboratory Systems, 117, 116-128, 2012.

(14d) Multivariate Latent Variable Modeling Across the Lifecycle of a Product

Garcia-Munoz, S., Pfizer Worldwide R&D
Muteki, K., Pfizer Inc
Originally presented on: 11/3/2013 17:15:00 - 17:40:00

This talk focuses on the use of multivariate methods (such as PCA ,PLS are other advanced derivations of these) across the lifecycle of a new product. For the cases presented here, the lifecycle of the new product begins at the design phase with the selection of raw materials. This selection can be simultaneously done with the design of the process train, along with its processing conditions. Another component of the design phase is the development of analytical techniques to detect the concentration of a component, or to characterize the end product. These methods (ranging from the spectroscopy based to more advanced digital-image based or multi-signal based techniques) are critical in the development of monitoring, real time release and control schemes. As the product evolves, multivariate tools are applied in the transfer of the product from laboratory and pilot scale to commercial manufacturing to ensure the quality and properties of the end product are unchanged from what was produced at smaller scales. Once the product enters full scale manufacturing, multivariate latent variable methods are applied in monitoring of the state of the process and in control algorithms that take advantage of the degrees of freedom in the process to compensate for disturbances entering the system.

(14e) Dealing With Missing Data and Synchronization Issues in Batch Process Monitoring

Ferrer, A., Polytechnic University of Valencia

Originally presented on: 11/3/2013 17:40:00 - 18:05:00

After visiting professor John F. MacGregor in Fall 1997 I got really fascinated by the potential of multivariate latent variable methods for batch process monitoring. Based on the inspiring contributions of John I went on studying new methods to ease and improve batch process monitoring.

In this talk I will present some of our novel contributions in this field, based on previous John´s work. In particular I will review two issues: missing data and real-time batch synchronization.

In process chemometrics missing data can appear for a number of reasons: failures in sensors or in the communication between the instrumentation and the digital control system (DCS), sensors taken offline for routine maintenance, manual samples not collected at the required times, data discarded by gross measurement errors, and sensors with different sampling periods. In real-time-batch process monitoring and control, at each time t the future behaviour of the process trajectories is unknown, and then can be treated as missing values. Out of the proposed methods, Trimmed Scores Regression (TSR) [1] turns out to be a simple an efficient method to estimate missing data.

Usually batch pace evolution is different batch-to-batch yielding misaligned key process events. Real-time batch synchronization is a critical issue for real-time batch process monitoring. For this purpose the Relaxed-Greedy Dynamic Time Warping (RGTW) algorithm [2] will be introduced.

[1] Arteaga F., Ferrer A. Dealing with missing data in MSPC: several methods, different interpretations, some examples. J. Chemometrics 2002; 16: 408–418.

[2] González-Martínez J., Ferrer A., Westerhuis J. Real-time synchronization of batch trajectories for on-line multivariate statistical process control using dynamic time warping, Chemometrics and Intelligent Laboratory Systems 2011; 105: 195-206.

(14f) Some Perspectives On the Impact of BIG Data On Process Systems Engineering
MacGregor, J., Dept. of Chemical Engineering, McMaster University
Originally presented on: 11/3/2013 18:05:00 - 18:30:00

This presentation offers some advice to young researchers based on my own experiences in industry and academia. It also provides a personal perspective on the impact that BIG data is having on PSE. The past has seen the rise of multivariate latent variable methods to treat many large messy data problems, but the data sets continue to get bigger in almost every area.

Control and optimization have almost invariably been based around the use of theoretical models and or structured empirical models fit to plant data. However, what if many informative measurements such as color images, vibrational spectra, etc. are available to help define the state of the process, or what if the appearance of a product obtained from an image is the objective of the optimization? A general conclusion might be that, as we continue to collect more and more information through more advanced sensors, our traditional PSE reliance on fundamental or structured empirical models becomes less and less viable and hybrid approaches are needed.

(39a) Towards Personalized Therapy. Biomolecular Complexities Beyond Gene Expression

Ramkrishna, D., Purdue University

Originally presented on: 11/4/2013 8:30:00 - 9:05:00

Treatment of cancer has been a perennial challenge to clinicians and researchers alike for more than a century. Existing treatment modalities are found to be effective only for patients of select genetic and phenotypic backgrounds. However, large genetic and phenotypic variations exist among patients so that the ***current approach of 'standard-dose-for-all' causes severe toxicity in some patients and insufficient treatment in others***.

Personalized medicine has been in practice in some form or other ever since medicine has been in existence. With the discovery of new complex diseases, increased incidence and mortality rates of existing diseases, the emphasis on personalized medicine has been gaining momentum in recent decades. The crux of personalized medicine lies in identifying genetic and biomolecular traits specific to a patient, followed by quantitative clinical decision making. With the mapping of the human genome and recent developments in 'omic' technologies, a deluge of patient-specific data have become available which can be used to characterize an individual patient to some extent so as to predict specific response to a drug. The ease of simultaneous characterization of several thousand gene expressions has greatly enhanced the capabilities of pharmacogenomics. Yet, the growth of personalized medicine has been far from optimum due to what we believe to be skewed application of these latest technological advancements. In many instances, pharmacogenomics is being interchangeably used for personalized medicine, which in reality is only *one* of the many facets of personalized medicine. The expression/observation of the molecular phenotype (e.g. proteins, small molecules) from the gene sequence involves numerous transcriptional and translational processes. Furthermore, molecular phenotypes act on various cell populations through several processes to produce clinical phenotype. All of these processes are stochastic and temporal in nature resulting in ***completely different response profiles for patients with the same genotype***. A paradigm shift from evidence-based medicine to mechanism-based medicine will ensure that each patient is treated according to his/her own mechanism and thus should be the impetus for expanding implementation and utilization of personalized medicine.

Under these circumstances, a holistic approach, *integrating* many levels of biomolecular entities and events, would provide more relevant information for clinical decision making. Given the complexity of biological processes and the amount of available data, it would be unreasonable to expect a simple deductive process or empirical models to guide personalized treatment. Quantitative models, derived from first principles and suitably empowered by systems theoretic methodology, have had a history of demonstrated successes in many scientific and industrial applications. This presentation will attempt to show that mathematical models have the potential to ***identify functionally interrelated sets of reactions and metabolites that are causally related to diverse pathophysiological conditions***. It will further deliberate into the path to personalized medicine towards the development of a decision-support mechanism for physicians to quantitatively predict patient response and adjust dosage for an individual. Such a quantitative approach has the potential to greatly enhance the decision making capabilities of the treating physicians and improve the quality-of-life among cancer patients.

(39b) A Perspective On the Use of Mechanistic Models Alongside Statistical Design of Experiments to Define a Design Space in Drug Substance

Tom, J. W., Bristol-Myers Squibb

Originally presented on: 11/4/2013 9:05:00 - 9:40:00

In Chemical Process Development for small molecule drug substance, the design of an optimal synthesis and manufacturing process must meet key elements of safety, product quality, efficiency, robustness, economics, intellectual property constraints and greenness/sustainability. Defining a design space, in which to operate the process parameters across the sequence of unit operations to meet the design goals, encompasses the work of many chemists and chemical engineers through a range of activities (lab experimentation, data analysis, modeling, and plant scale-up). The resulting data package and design space form the basis for the eventual Chemistry, Manufacturing and Controls section of regulatory filings for New Drug Applications.

The Quality by Design approach emphasized by the FDA in the last decade and the compression of timelines to develop a drug candidate provides a strong need for efficient process development and rigorous scientifically-based approaches to develop this design space. Elements of this approach encompass high-throughput experimentation, modeling (both mechanistic and empirical), data visualization, statistical design of experiments, and appropriate scale-up verification. This talk will provide a perspective on the approach at Bristol-Myers Squibb incorporating these elements.

(39c) The Impact of Modeling in Drug Product and Process Development

Hilden, J., Eli Lilly and Company

Originally presented on: 11/4/2013 9:40:00 - 10:15:00

In most large pharmaceutical companies, the commercial drug product and process development cycle begins in earnest after the first dose has been administered to humans and ends when the manufacturing process has been developed, scaled up, and transferred to the intended manufacturing site. Before First Human Dose (FHD), the goal of development is to get the Active Pharmaceutical Ingredient (API) into the blood stream of a healthy volunteer. After successful FHD, the goals change, and a commercially viable manufacturing process must be identified to reliably produce large quantities of stable drug product at a reasonable cost. The input to the development engine is the API, usually as a white crystalline powder. The output is knowledge: a recipe for robust manufacturing of tablets, capsules, or other drug products. Formulation scientists and process engineers who develop this recipe must reduce the enormous design space of possible materials and unit operations into a single recommended path forward. It is in these decisions that a judicious use of modeling and scientific planning can frequently reap significant rewards. In this work, a review of relevant modeling approaches is provided. Those areas of significant impact as well as areas of significant need are highlighted.

(39d) Application of Single Use Technologies in Large Molecule Development and Manufacture and the Consequences for Large Molecule QbD

Frank, G., Amgen

Originally presented on: 11/4/2013 10:15:00 - 10:50:00

To be expanded shortly.

You have spoken several times in recent years about the single use revolution and this would be an interesting opportunity to expand on that for the community. On the other hand if there is a more burning and general topic in the large molecule field that you would like to address instead, that is also an option.

You have spoken several times in recent years about the single use revolution and this would be an interesting opportunity to expand on that for the community. On the other hand if there is a more burning and general topic in the large molecule field that you would like to address instead, that is also an option.

You have spoken several times in recent years about the single use revolution and this would be an interesting opportunity to expand on that for the community. On the other hand if there is a more burning and general topic in the large molecule field that you would like to address instead, that is also an option.

(144a) Tales From the Trail: A Continuing QbD Journey

Lepore, J., Merck & Co., Inc.

Originally presented on: 11/4/2013 12:30:00 - 12:49:00

Tales from the Trail: A Continuing QbD Journey

Quality by Design has transformed the way in which many companies approach process development. By starting with patient needs and following a structured framework that follows the voice of the patient into an assessment of risks that may inhibit meeting those needs, then establishing process deliverables and controls, it is generally accepted that processes are better designed and are more robust as they enter commercial production. Although it is recognized that QbD is the right thing to do for patients and customers, many challenges and barriers remain to achieving a landscape in which all products are developed and/or communicated in that framework. For example, the lack of global harmonization, of aligned perception of residual risk, and of interpretation of ICH guidance and supporting materials, have made it increasingly difficult to navigate the global regulatory process with a QbD submission.

This presentation will provide a case study, and outline some ways to drive QbD forward from its current crossroads.

(144b) Quality By Design Pilot to Reality: The Honeymoon Phase
am Ende, M. T., Pfizer Worldwide Research and Development
Originally presented on: 11/4/2013 12:49:00 - 13:08:00

Chantix® was Pfizer's first product filed using the Quality by Design approach, and was Pfizer's first product to participate in the FDA's Pilot Program. The drug product was approved 9 months after regulatory submission on May 10, 2006. Chantix® is available as 0.5 and 1.0 mg immediate release tablets for the treatment of smoking cessation. With a successful launch campaign, the product demand exceeded the original forecasts. The Pfizer Global Manufacturing site in Illertissen, Germany was challenged to streamline processing to maximize production rates. Through application Quality by Design and lifecycle efficiency improvements, PGM-Illertissen Right First Time team set out to do just that! In this presentation, I will demonstrate the flexibility and benefits fostered by the QbD approach. The manufacturing process description and metrics summarizing changes that were used to streamline the process will be contrasted to recent experience in C. Seymour's talk entitled "Quality by Design Pilot to Reality: The Stormy Years".

(144c) Quality By Design Pilot To Reality: The Stormy Years

Seymour, C., Pfizer Inc
Am Ende, M. T., Pfizer Global Research and Development
Originally presented on: 11/4/2013 13:08:00 - 13:27:00

This presentation compliments **Quality by Design Pilot to Reality: The Honeymoon Phase talk** by M. am Ende which overview the early QbD experiences and results. **Quality by Design Pilot to Reality: The Stormy Years** will provide an overview of several recent commercial applications and compare the current QbD experiences with, non QbD reviews and the pilot experience.

(144d) Road to the Promised Land of QbD: Informatics and Modeling Challenges

Venkatasubramanian, V., Columbia University

Originally presented on: 11/4/2013 13:27:00 - 13:46:00

Developing and manufacturing pharmaceutical products in a timely manner while ensuring quality is a complex process that requires handling vast amounts of varied data, information, and knowledge. The new FDA framework of PAT-QbD, and the ICH Q8R, Q9, and Q10 (step 2) guidelines, pose major modeling and informatics challenges that have not been fully appreciated by the pharmaceutical industry, and will require radically new approaches and significant investments. However, recent progress in information and knowledge management technologies offer hope. One such approach is called *Ontological Informatics*. An ontology is a formal and explicit specification of a shared abstract model of a phenomenon through identification of its relevant concepts. An ontology defines and semantically describes data, information, and models. In this talk, I will discuss the use of such a framework for modeling two different types of knowledge often needed for QbD: mathematical knowledge, as in equations, and decision-making knowledge, such as decision trees and heuristics. I will discuss the challenges, opportunities and emerging trends using case studies from pharmaceutical manufacturing.

(144e) Process Understanding in the Context of QbD: Current Approaches and Desired State

Khinast, J. G., Graz University of Technology

Originally presented on: 11/4/2013 13:46:00 - 14:05:00

The development of process understanding is a critical element of the quality design framework. In this context, designed experiments are currently considered the best method for developing process understanding, mostly via MVDA models. However these models lack extrapolatory capabilities, are scale-dependent and may miss critical system behavior. Thus, novel approaches are required that rely on mechanistic understanding of the processes. Several methods, which also can be used for rational risk ranking, are presented in this talk. Moreover, their respective advantages over the current methods are discussed.

In the presentation a model is presented of how detailed high-fidelity computer simulations can be used to support a quantitative risk evaluation process and how a scale-dependent design space can be transformed for routine manufacturing. Specifically, the coating of tablets and the mixing of powders are considered as model processes. For the purpose of these simulations advanced DEM and CFD tools are applied, in combination with advanced material characterization tools.

(144f) Quality By Design for Drug Substance Development - An Industry Perspective
Tummala, S., Bristol-Myers Squibb
Originally presented on: 11/4/2013 14:05:00 - 14:24:00

The principles that embody the Quality by Design paradigm have been increasingly adopted by pharmaceutical companies over the past number of years. Recent ICH guidelines and health authority opinions have contributed to the formalization of these principles during development, manufacturing and in regulatory submissions. While it is clear that application of QbD principles has definitely led to increased process understanding and buying down of risk, there remain significant challenges in adopting this approach universally for regulatory filings. Through some case-studies, the benefits of adopting this approach to drug substance development and the challenges that remain will be discussed

(144g) Quality be Design and Regulatory Science: An Academic Perspective
Wang, H. Y.,
Originally presented on: 11/4/2013 14:24:00 - 14:43:00

. In advanced biomedical manufacturing, Quality System Regulation (QSR) has already been used by CDRH in medical device regulations and oversight since 1997 while Quality By Design (QbD) initiative by FDA has recently been adopted in pharmaceutical and biopharmaceutical sectors. FDA defines QbD as doing the science to design process performance characteristics, not merely empirically derived from the performance of traditional test batches. The specific challenge in QbD has been on how to obtain essential scientific understanding of intrinsic cause-effect relationships between process and products for improving process control and reducing products risk and eventually reaching the desired state of potential real time product release of any biomedical product with minimal final quality testing. The slow adoption of QbD can partly be traced to our current life science science and engineering research, education and training. Many academic research groups have been investigating unique and challenging biomedical products and technologies through advances in genomic and biomedical research. Often, the potential product candidates coming out from the lab are very complex in nature that are often ill defined and require custom-designed formulations, distribution and delivery that post challenging scale-up and other technical and regulatory issues. Unfortunately, there is nearly a complete absence of any focus on manufacturing science related efforts that emphasize design and quality controls, data reproducibility and information traceability. This is surprising, given that reliable and cost effective manufacturing of biomedical products and devices have been identified to be a major roadblock for many innovative biomedical solutions. We shall discuss these roadblocks and advocate some possible solutions to remedy this obvious deficiency.

(144h) Why Are We So Risk Adverse?

Fermier, A., Janssen Pharmaceuticals
Weber, J., PDMS
Cunningham, J., PDMS
Mehrman, S., PDMS
Schaefer, E., Janssen Pharmaceutical Companies of Johnson and Johnson
Barnthouse, K.,
Harkins, A., Janssen Pharmaceuticals
Hanna, B.,
Janovjak, M.,
Cedeno, W., Janssen Pharmaceuticals
Originally presented on: 11/4/2013 14:43:00 - 15:02:00

Quality by Design is a great concept to drive corporations towards a more streamlined and disciplined mechanism of developing new medicines. However, it has been a long drawn out conversation with specific case studies, and achieving a sustainable change in this direction is proving difficult. Is it because we are risk adverse? Obviously not in our industry with failure rates of molecules reaching >80% from Phase 1 to launch. We take risks in what we move to the clinic, but how we design and approach our process development one could argue is still very conservative. The rationale for the conservative approach is the complexity of the manufacturing process itself and substantial consequences of failing ultimately the patient that drives the conservative behavior. In this presentation we will be discussing our risk based approach we are implementing, our challenges and overall strategy to overcome to help ensure the movement to the risk based approach is a sustainable change.

(178a) A Novel Computational Oral Absorption Simulation Tool to Integrate API Crystal Properties and Drug Product Design Attributes to in Vitro and in Vivo Performance

Sagawa, K.,
Shanker, R. M., Pfizer Worldwide R&D
Li, R.,
Abbou Oucherif, K., New Mexico Institute of Mining and Technology
Braido, D., Rutgers University
Calado, F., Process Systems Enterprise
Bermingham, S. K., Process Systems Enterprise
Originally presented on: 11/4/2013 15:15:00 - 15:35:00

Purpose: Develop a Systems-based Pharmaceutics holistic modeling approach to integrate API crystal properties and drug product design attributes to in vitro and in vivo performance .

Methods: The gPROMS modeling platform is now used to implement gCOAS (**C**omputational **O**ral **A**bsorption **S**imulation) a novel computational oral absorption modeling tool, with inherent capability for integration with other products available in the same platform.

Results: gCOAS model has several advanced features for simulating oral absorption. Features include mass and charge balances of drug species (ionic, non-ionic and micellar) and physiological ions in solution in the gastrointestinal (GI) tract. A population balance approach is used to represent the particle size distribution of each solid phase either present in the drug product or formed during transit through the GI tract. Advanced features of nucleation and crystal growth kinetics include calculating the supersaturation with respect to the relevant species forming the solid phase. Drug dissolution is modeled incorporating surface pH of solids, changes in pH of GI tract, and speciation of drug in solution, including solubilization into bile micelles. The GI tract is modeled based on anatomical segments of which the user can choose to implement any number of sub-segments . Each sub-segment is defined by a residual volume, pH, bile concentration etc. Additionally all incoming fluids including secretions based on human physiology of GI tract is modeled via rapid re-absorption from the GI wall to attain steady state net water volume.

(178b) Differential Diagnosis of Primary Hyperparathyroidism and Extent of Progression Through Short-Term Calcium Loading

Christie, C. R., Virginia Polytechnic Institute and State University
Achenie, L. E. K., Virginia Polytechnic Institute and State University
Ogunnaike, B. A., University of Delaware
Originally presented on: 11/4/2013 15:35:00 - 15:55:00

In the human body, total plasma calcium is maintained within a specific range (8.5-10.3 mg/dl) [1] via a regulatory system that employs parathyroid hormone (PTH) and calcitriol (CTL) to moderate Ca flux between the plasma and each of the organs: kidneys, intestines and bones. Disorders in the regulatory organs cause increased hormonal activity which subsequently leads to chronic imbalances in plasma Ca levels. For example, enlarged parathyroid glands (PTG) associated with primary hyperparathyroidism (PHPT) cause over secretion of PTH which leads to high plasma Ca levels (hypercalcemia)[2].

A common feature in the major pathologies associated with hypercalcemia is the excess secretion of PTH and/or PTH-related protein, leading to increased plasma Ca. The three most common pathologies, accounting for 90% of hypercalcemic cases are Familial Benign Hypercalcemia (FBH), Humoral Hypercalcemia of Malignancy (HHM) and PHPT. FBH is a missense mutation of the calcium-sensing receptor, located on the surface of the parathyroid cells, causing increased PTH secretion. HHM is caused by cancerous tumors producing PTH-related protein (PTHrP). PHPT is often confused with FBH and/or HHM during the early stages of progression because the three pathologies have an identical effect on Ca-related biochemistries. A series of biochemical tests and family history checks are usually necessary to differentiate among these diseases [2].

Previously, we used an engineering control system framework to represent physiological plasma Ca regulation [3]. This framework describes the dynamics of PTH from the PTG, calcitriol from the kidneys, bone cells (osteoblasts and osteoclasts) and plasma calcium. These are mapped onto the controller, the actuators and the process. Following validation of the overall model with published clinical data we successfully simulated Ca-related pathologies through induced defects in the control system components (sensor, controller and/or actuator).

In this presentation, we provide a systematic approach for differentiating among FBH, HHM and PHPT based on the unique hormone/ion responses to short-term Ca disturbance in each pathology model. Additionally, based on the changes in intrinsic parameters associated with PTG behavior, we show that the extent of PHPT progression can be reasonably predicted and the enlarged gland size estimated *a priori.*

We hypothesize that the proposed disease diagnosis strategy would be useful for (a) early detection of PHPT in at-risk or asymptomatic patients and, (b) differential diagnosis of FBH, HHM and PHPT where the pathologies have identical presentations and are difficult to distinguish using existing clinical methods.

1. Kovacs, W.J. and S.R. Ojeda, *Textbook of Endocrine Physiology*. 6th ed2012, Oxford: Oxford University Press. xii, 462 p.

2. Fraser, W.D., *Hyperparathyroidism.* Lancet, 2009. **374**(9684): p. 145-58.

3. Christie, C.R., L.E.K. Achenie, and B.A. Ogunnaike, *A Control Engineering Perspective to Modeling Calcium Regulation and Related Pathologies*, in *Annual Meeting October 28- November 2,*2012.

(178c) Polymer Inhibition of Crystal Growth for Improved Drug Solubility

Schram, C.,
Beaudoin, S., Purdue University
Taylor, L., Purdue University
Originally presented on: 11/4/2013 15:55:00 - 16:15:00

A large percentage of active pharmaceutical ingredients (APIs) in the drug development pipeline demonstrate poor aqueous solubility as crystalline solid forms. This poses a problem for oral therapeutics due to decreased bioavailability. Utilizing the amorphous form of the drug rather than the crystalline form is an effective method to alleviate solubility issues, due to its greater overall free energy. Much emphasis has been placed on methods to stabilize the inherently unstable amorphous form and prevent crystallization. One such method utilizes polymers to act as a mechanical barrier, occupying growth sites. The effectiveness of this method, however, varies for each polymer-drug combination.

In this work, we study polymer inhibition of crystal growth of the drug felodipine in solution. We use a rotating disc apparatus (RDA) to eliminate diffusion effects, which allows us to study growth kinetics controlled purely by integration at growth sites. We then use atomic force microscopy (AFM) to quantify the amount of adsorbed polymer, as well as to document its adsorbed conformation on solid felodipine. These results reveal insights into the mechanisms by which polymers can effectively inhibit crystal growth.

(178d) Experiences in the Corporate-Wide Deployment of Advanced Modeling Technology: Making the Complicated Look Simple, But No Simpler Than That!

Garcia-Munoz, S., Pfizer Worldwide R&D
Schmitz, P., Pfizer Worldwide R & D
Originally presented on: 11/4/2013 16:15:00 - 16:35:00

The development of modeling technology in an organization is typically in the hands of a small sub-group of subject matter experts. Once these models are mature enough, a natural step is the deployment of such models for the general audiences, who will in turn use these models as a tool to their job responsibilities.

In the worst of the cases, the deployment process for a model will involve training the end-user to manipulate the native tool in which the model was built on (e.g. gPROMS, Aspen Plus, Dynochem, Fluent, MATLAB, etc). This training process can be challenging due to the diversity of end-users in a large corporation. This also implies the need to install the native tool in the end-user's computer and the potential consumption of an expensive license for the rather simple "use of a model" rather than the more challenging "development a model". In better cases the deployment process with involve the development of interfaces built using event-driven languages like Visual Basic.

Either of these scenarios involves many complexities: *i*) the establishment of a mechanism such that the end user has access to the latest version of the model, *ii*) the proper training of the user in the native tool, *iii*) the sub-optimal use of expensive licenses, *iv*) the additional need of a framework to store/archive results from the use of the models, *v*) the need to maintain ad-hoc interfaces, among others.

In this talk we share our experiences implementing a web-based solution to achieve the corporate-wide deployment of models trying to solve some of the complexities before mentioned. Our talk addresses the necessary protocols and efforts involved in launching and maintenance of such a framework, including the different levels of documentation, the efforts to gain end-user participation and to properly manage the lifecycle of a model will be discussed.

Such a framework has also served as a trigger for the broader discussion from the business lines, as to where in the development cycle can (and should) these tools be applied, and to what level should decisions be made with the results from in-silico technology. Widening the audiences that can now access complex computational solutions has also brought new challenges regarding data-transfer (especially at the interface between discovery and development). It has also sparked the dialog between different communities of "modelers" that have been doing very different types of "modeling" of some form or another and the need to establish a system that addresses the multiple needs of a very diverse community of subject matter experts. We will also touch upon the ways this framework has changed the way the subject matter experts and the informatics personnel exchange information and respond to the needs of the end user. Our intention is to spark the debate about the different ways such a platform could potentially change the form we transfer or share knowledge from company-to-company, company-to-vendor, company-to-academia, and company-to-agency.

(178e) Study of GW786034 Crystallization Kinetics in Pazopanib Eye Drop

Dai, W., GlaxoSmithKline

Originally presented on: 11/4/2013 16:45:00 - 17:05:00

Age-related macular degeneration (AMD) is a leading cause of blindness and visual impairment in older adults (>50 years). It occurs in "dry" and "wet" forms. The majority of significant vision loss is caused by wet AMD, which is attributed to the abnormal blood vessel growth in the choriocapillaris of eyes. The current standard care is repeated injections of anti-VEGF agents in the eyes. GSK has been developing Pazopanib eye drop (a multi-targeted receptor tyrosine kinase inhibitor) as an alternative treatment for wet AMD. Formulations under development are highly supersaturated Captisol based GW786034 solution. This presentation will describe our approach taken to understand long term stability of the eye drop formulation (drug product) through GW786034 crystallization kinetics study, in which concentrations of GW786034 and Captsol, pH and temperature were examined and PAT was employed. Mechanism of GW786034 solid formation and impacts of supersaturation of the formulations on drug product manufacture and storage will also be discussed.

(178f) Quality By Design in Integrated Processes: From Drug Substance to Drug Product

Gil, M. Sr., Hovione Farmaciencia
Gaspar, F. Sr.,
Matos, N., Hovione
Originally presented on: 11/4/2013 17:05:00 - 17:25:00

This paper will present a methodology that can be used when applying Quality by Design (QbD) in consecutive process steps, namely when integrating drug substance and drug product processes. It will show how process parameters and material attributes can be effectively combined expanding the overall design space and hence process flexibility in both cases. This work will also present a comparative study between similar processes using Quality by Design and Traditional approaches. Process metrics and performance indicators have been measured after extensive exposure of both processes to manufacturing campaigns. QbD based process showed for all parameters a significant improvement regarding the traditional approach. The higher development costs of the QbD based approach were rapidly covered by the improved performance at commercial manufacturing stage.

(178g) Concomitant Co-Crystallization On Engineered Surfaces

Shi, X.,
Wong, S. Y., Massachusetts Institute of Technology
Yang, X.,
Myerson, A. S., Massachusetts Institute of Technology
Originally presented on: 11/4/2013 17:25:00 - 17:45:00

Concomitant Co-Crystallization on Engineered Surfaces

Xingyi Shi, Shin Yee Wong, Xiaochuan Yang and Allan S. Myerson

Department of Chemical Engineering, Massachusetts Institute of Technology, 77 Massachusetts Avenue, 66-568, Cambridge, Massachusetts 02139, United States

Co-crystals are crystalline materials made up of two or more components. In pharmaceutical systems, co-crystals are used to enhance the physical properties (e.g. solubility or melting point) of a drug substance via non-covalent interactions with a co-former. Numerous high throughput screening methods have been proposed[1]. However, few of these studies examine polymorphism of the co-crystals. In our previous work[2], patterned self-assembled monolayers (SAMs) were shown to be an attractive tool for polymorph screening. To expand the application to co-crystals, the patterned SAMs substrates were used for the concomitant co-crystallization of caffeine-oxalic acid.

Bi-functional patterned SAMs were prepared by photolithography using silicon wafers as the base substrate coated with 5 nm titanium and 50 nm gold layers. The patterned gold substrates have an island diameter of 500 Âµm, with the hydrophilic gold island functionalized with the ?COOH group, and hydrophobic long-chain alkyl group on the silicon substrate. Co-crystallization experiments were conducted by forming an array of droplets of caffeine and oxalic acid solution (2:1 molar ratio) in chloroform: methanol (7:2 v/v) onto the SAMs. Then, the solvents were allowed to evaporate. Once the co-crystal is formed, the crystals were characterized by Raman spectroscopy and X-ray diffraction.Â Compared to the Raman spectrum of the co-crystal (reference) reported earlier[3], polymorphs with distinct Raman peaks at 1076 and 1588 cm^{-1} were observed. With six repeated experiments, more than 50% of the islands were found to be the reference co-crystal. Â On non-reference co-crystal islands, majority of the crystals have peaks at 1076 and 1588 cm^{-1} in addition to the co-crystal reference peaks, with some of them having a lone peak at either 1076 or 1588 cm^{-1}. Besides, the XRD diffraction pattern also showed additional peaks at 2θ of 30° and 32.3°, further confirming the presence of a new co-crystal form.

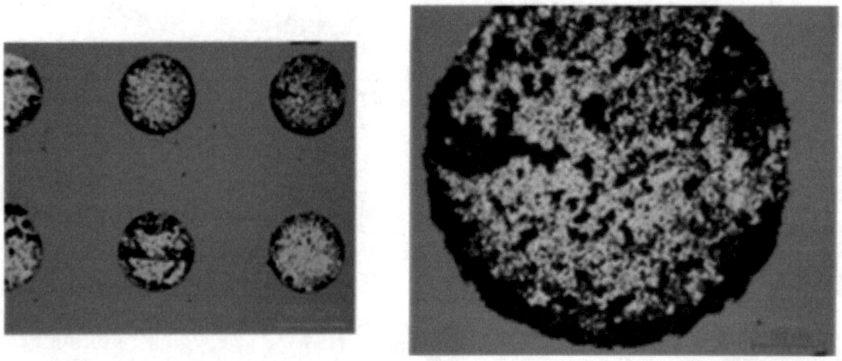

Figure 1. Co-crystals grown on SAMs at increasing magnification

Reference:

1. Elbagerma, M. A.; Edwards, H. G. M.; Munshi, T.; Hargreaves, M. D.; Matousek, P.; Scowen, I. J. *Crystal Growth & Design* **2010**, *5,* 2360-2371.

2. Singh, A.; Lee, I. S.; Myerson, A. S. *Crystal Growth & Design* **2009**, *2,* 1182-1185.

3. Trask, A. V.; Motherwell, W. D. S.; Jones, W. *Crystal Growth & Design* **2005**, *3,* 1013-1021.

(194a) Nanoparticle-Assisted, Mechanism-Based Combination Photodynamic and Irinotecan Therapy for Pancreatic Cancer

Huang, H. C., Massachusetts General Hospital, Harvard Medical School
Mallidi, S.,
Mai, Z.,
Goldschmidt, R.,
Rizvi, I.,
Liu, J., Massachusetts General Hospital, Harvard Medical School
Chiang, C. T.,
Hasan, T.,
Originally presented on: 11/4/2013 15:15:00 - 15:35:00

The purpose of this study is to strategically combine two clinical-relevant, nanotechnology-based therapies to facilitate rapid clinical translation and immediately improve on the dismal statistics of pancreatic cancer (PanCa) patients. We hypothesized that benzoporphyrin derivative (BPD)-based photodynamic therapy (PDT) (Phase I/II study, solid PanCa) destroys tumor efflux transporters, which may help maintain high intracellular concentrations of Irinotecan (CPT-11) (Phase III study, metastatic PanCa) to reduce tumor burden and prolong survival. We test our hypothesis in orthotopic PanCa models.

Two types of liposomes were fabricated by adapting procedures from literature. They are: (i) Liposome with BPD in lipid bilayer (LBPD) and (ii) Liposome encapsulating CPT-11 in aqueous core (LCPT-11). Lipids (DPPC, DOTAP, Cholesterol, DSPE-mPEG at a molar ratio of 2:0.2:1.0:0.2) were mixed in chloroform (for LBPD, dissolve with 0.2 mM BPD), and the chloroform was evaporated. Lipid films were rehydrated for 2h in an aqueous solution (for LCPT-11, contain 7 mM CPT-11) with freeze thaw cycles. The resulting dispersion was extruded through polycarbonate membranes (100 nm pore size) to form unilamellar vesicles. Liposome size and polydispersity were measured by dynamic light scattering. BPD (or CPT-11) concentration was determined by UV-Vis spectroscopy. Human pancreatic cancer cells (MIA PaCa-2 or AsPC-1, ATCC) were implanted orthotopically in Swiss Nu/Nu mice (4-6 weeks old, ~25 mg) on day 0. Animals were anesthetized with Ketamine/Xylazine and the pancreas was exteriorized. Cells (1 x 10^6 in 50 μL of Matrigel-containing media) were injected into the pancreas using a 301/2-gauge needle, and the incision was closed with 4-0 sutures. Treatments were initiated when the tumors reach ~35 mm^3 on day 9 (determined by ultrasound imaging). Tumor bearing mice were intravenously (tail vein) injected with LBPD (0.25 mg/kg) and LCPT-11 (20 mg/kg) 1h before light administration. Interstitial PDT (690 nm laser, 100mW/cm^2, 75 J/cm^2) was performed on the exteriorized pancreas of the anesthetized animal, and then followed by wound closure with sutures.

We have delivered reproducible LBPDs (137±9 nm) and LCPT-11s (122±5 nm), with a polydispersity index less than 0.05, were found stable for up to 3 weeks of storage. The BPD and CPT-11 loading efficiency in liposomes were found to be ~75% and ~50%, respectively. The longitudinal ultrasound monitoring of orthotopic tumor volume in response to combination LBPD-PDT and LCPT- 11 was carried out with appropriate controls. We observed that LBPD-PDT significantly enhances the tumoricidal efficacy of LCPT-11 and significantly inhibited tumor growth up to at least 3 weeks post-treatment ($p < 0.05$). Tumor volumes for the combination group on day 30 were ~3 fold and ~5 fold less than single treatments and no treatment groups, respectively.

Photodynamic therapy (and Irinotecan chemotherapy) alone has already shown promise in treatment of PanCa in clinical studies. This study recognizes that the genetic complexity and heterogeneity of PanCa make it extremely difficult for any single treatment to impact outcome. To overcome therapy resistance, it is critical to understand the limitations of single treatment and develop new combination regimens based on interactive mechanisms. We performed pilot studies in orthotopic PanCa models that demonstrated LBPD-PDT could improve the efficacy of liposomal Irinotecan treatment. We anticipate the findings of this study, based on two clinically relevant treatments, will form the basis for rapid translation of a novel combination regimen for PanCa patients.

(194b) Structure of Myoglobin Confined in Nanoporous Silica

Siefker, J., Rensselaer Polytechnic Institute
Krutyeva, M.,
Biehl, R.,
Coppens, M. O., Rensselaer Polytechnic Institute
Originally presented on: 11/4/2013 15:35:00 - 15:55:00

There is considerable interest in the use of protein immobilization in porous hosts to address challenges in biotechnology and therapeutics. Hence, there is a need to develop a more thorough understanding of the fundamental behavior of proteins in confinement. Pore surface properties such as charge or hydrophobicity are known to dramatically affect the behavior and structure of interacting proteins.[1,2] Our studies of confinement effects in nanoporous particles has shown that geometric properties such as surface curvature play a significant role, and pores of high curvature may stabilize the native protein structure.[2] This could protect enzymes from extreme environmental conditions.[3] While these results are promising, they only paint a partial picture of the structure and behavior of confined proteins, and need to be used in conjunction with other techniques to provide a clearer molecular understanding. This investigation attempts to further the understanding of a model system – myoglobin in nanoporous silica SBA-15 – by determining the overall structure and physicochemical properties of confined myoglobin. Structural information is obtained through the use of direct probing techniques such as small angle neutron scattering (SANS) to observe whole protein structural parameters, paired with previously obtained FTIR secondary structural information,[3] and using these parameters in a geometric model that evaluates the packing of globular proteins, modeled as ellipsoids, in cylindrical channels.[4] The information obtained in this study provides further insight into the fundamentals of protein confinement.

References:

1. Yuan L, Yu Q, Li D, Chen H. Surface Modification to Control Protein/Surface Interactions. Macromolecular Bioscience. 2011;11(8):1031–1040. doi:10.1002/mabi.201000464.
2. Sang L-C, Coppens M-O. Effects of surface curvature and surface chemistry on the structure and activity of proteins adsorbed in nanopores. Physical Chemistry Chemical Physics. 2011;13(14):6689. doi:10.1039/c0cp02273j.
3. Lynch M., Siefker J., Coppens M-O. Confinement Protection Effects of Mesoporous Silica SBA-15 On Myoglobin, in an Environment Inspired by the Groel/ES Chaperonin System. AIChE Annual Meeting. 2012.
4. Sang L-C, Vinu A, Coppens M-O. General Description of the Adsorption of Proteins at Their Iso-electric Point in Nanoporous Materials. Langmuir. 2011;27(22):13828–13837. doi:10.1021/la202907f.

(194c) Investigation of the Drug Distribution Within Nanoparticles By Transmission Electron Microscopy (TEM)

Gjersee, T.,
Zhan, X., University of Washington
Yacoob, C., University of Washington
Wang, L., University of Washington
Shen, H., University of Washington
Originally presented on: 11/4/2013 15:55:00 - 16:15:00

The drug distribution pattern within particles are critical for controlling the release of the drug from particles. Due to the effect of mixing within particles and at the interface, different fabrication processes can result in different patterns of drug distribution. In this study, we investigated two fabrication processes, sonication- and microfluidic-enhanced nanoprecipitation. In order to visualize by TEM, quantum dots with 5 nm in diameter were used as model drugs. We examined the effect of sonication strength and the inlet configuration of microfluidic system on the distribution of quantum dots within a particle. The sonication method yielded more uniform distribution of quantumn dots within a particle than microfludic system. The inlet configuration of the microfluidic system has a significant impact on the distribution of quantum dots within the particles. In some configurations, quantum dots formed a cluster while others quantum dots were uniformly distributed throughout the particle. We are exploring the mixing with the particle and at the interface of two fluids in order to offer insights into the process design for desirable drug contribution within a particle.

(194d) An Ex Vivo Human Plaque Model for the Design of Amphiphilic Nanoparticle-Based Therapeutics for the Management of Atherosclerosis

Petersen, L. K., Iowa State University
York, A. W., New Jersey Center for Biomaterials
Zablocki, K. R., Rutgers University
Fan, C., Rutgers University
Lewis, D. R., Rutgers University
Uhrich, K. E., Rutgers University
Prud'homme, R. K., Princeton University
Haser, P. B., UMDNJ-Robert Wood Johnson Medical School
Graham, A. M., UMDNJ-Robert Wood Johnson Medical School
Moghe, P. V., Rutgers University
Originally presented on: 11/4/2013 16:15:00 - 16:35:00

Cardiovascular disease is the leading cause of death in the developed world, affecting more than 80 million people in the United States alone. This condition is largely caused by atherosclerosis, the formation of atheromatous lesions within the arterial walls, which is self-perpetuated through a low-level inflammatory response. The formation of atherosclerotic plaques is initiated by oxidation of low density lipoprotein (oxLDL) which triggers inflammation from endothelial cells. Monocytes are recruited to inflamed areas, which then differentiate into macrophages that can phagocytose resident oxLDL. OxLDL uptake via macrophages is unregulated, mediated by scavenger receptors (CD36 and SRA), and ultimately leads to foam cell formation and narrowing of the artery. Current strategies for treating this disease are limited in their ability to specifically delivery therapeutics to the site of lesions or lesion prone regions and cannot prevent further uncontrolled uptake of oxLDL. While statins prevent synthesis of LDL in the liver, they are unable to clear LDL introduced to the body via other pathways. Recently, our laboratories have identified a novel class of biomaterials containing a mucic acid backbone with lauryl side chains bound to polyethylene glycol (M12PEG), termed nanolipoblockers (NLBs). When formulated into micelles, these NLBs are able to bind to cells, macrophages and endothelial, via scavenger receptors and inhibit the uptake of oxLDL. They have also been shown to have a strong association with pre-inflamed cells, suggesting that these bioactive materials can accumulate at lesion sites. While the micellar formulation has strong therapeutic potential, serum containing conditions limit its bioactivity; thus, a more stable nanoparticle (NP) formulation has been established with superior bioactivity in serum containing conditions. NPs are fabricated via a flash nanoprecipitation process and contain the amphiphilic, M12PEG, shell and a hydrophobic core. This formulation enables the delivery of hydrophobic compounds with low water solubility to inflamed tissues.

In this work, we developed a novel *ex vivo* human carotid plaque model to evaluate bioactivity of a library of NPs comprised of a M12PEG amphiphilic shell and different hydrophobic cores. Hydrophobic cores investigated include the bioactive M12, anti-inflammatory vitamin E, and a combination of both bioactives. Using the *ex vivo* plaque model and primary human monocyte derived macrophages (hMDMs), bioactivity was assessed by evaluating the ability of the NPs to become internalized, inhibit oxLDL uptake, and reduce inflammation. The NPs were found to readily associate with hMDMs and the human plaques, even after a short 5 min incubation (similar exposure time to treatments administered via balloon catheter). This strong association within cellular regions of the plaque provides a feasible system for targeted drug delivery and enhanced therapeutic potential. All NP formulations demonstrated successful inhibition of oxLDL uptake at the site of lesions proving the therapeutic potential of these NPs. Additionally, the vitamin E and M12/vitamin E core NP formulations exhibited the lowest inflammatory gene expression, which were significantly less than that caused by oxLDL alone. In conclusion, this *ex vivo* human carotid plaque model has provided a novel method to test cardiovascular therapeutics on clinically relevant specimens. Employing this platform we have identified and optimized a novel NP-based therapeutic delivery system which will facilitate the management of inflammation and lipid accumulation at the site of atherosclerotic lesions.

(194e) Design of a Targeted Delivery System Based On Antioxidant-Loaded Biodegradable Nanoparticles for Neuroprotection in Cell Culture Model of Neurological Disease

Brenza, T., Iowa State University
Ghaisas, S.,
Vela-Ramirez, J., Iowa State University
Jia, F., Iowa State University
Anantharam, V.,
Kanthasamy, A.,
Narasimhan, B., Iowa State University
Originally presented on: 11/4/2013 16:35:00 - 16:55:00

In many neurodegenerative diseases such as Parkinson's (PD), Huntington's and Alzheimer's, there is a progressive loss of structure or function of neurons, including death of neurons.. The prevalence of these diseases is increasing along with the median age of our population. Mitochondrial dysfunction, oxidative and nitrative stress have been implicated in number of neurodegenerative diseases including PD. The existing treatment approaches for these diseases are symptomatic and fail to prevent the progression of the neurodegenerative process. Mitochondrial targeted antioxidants have been recently shown to protect against Parkinsonian toxicant, -methyl-4-phenylpyridinium (MPP^+) -induced production of reactive oxygen and nitrogen species production (ROS/RNS). Polyanhydride nanoparticles can be designed as a delivery system to elicit unique cellular responses such as particle internalization and directed intracellular trafficking. Through encapsulation of the payload within the polyanhydride particles, sustained release of these compounds can be achieved. Altering the polymer chemistry, functionalizing the particle surface, and/or controlling particle size enable the release of payload to targeted regions within the cell.

In this work we investigate the ability of biodegradable polyanhydride nanoparticles to enhance the delivery of mitochondrial targeted antioxidants to protect against MPP^+. The antioxidant utilized in these studies was apocyanin functionalized with a mitochondrial targeting ligand, mitoapocyanin. Using poly(sebacic acid) nanoparticles, we designed a delivery system for targeted release of mitoapocyanin to mitochondria of cells. The performance of the designed delivery system was evaluated for neuroprotection against MPP^+-induced cell death in both the MN9D dopaminergic cell line and mouse primary neurons. Cellular uptake of particles was evaluated by confocal microscopy of particles loaded with quantum dots. Intracellular trafficking of the particles to the mitochondria was evaluated by co-localization of the particle and mitochondria by confocal microscopy and confirmed by transmission electron microscopy.

(194f) Effect of Ultrasound and Emulsion On Release of Paclitaxel From Eliposomes

Javadi, M., Brigham Young University
Pitt, W. G., Brigham Young University
Originally presented on: 11/4/2013 16:55:00 - 17:15:00

ultrasound, the local pressure drops below the high vapor pressure that induces the formation and expansion of a vapor phase. The expansion ruptures the liposome, which releases the drug or genes locally. One of the advantages of emulsion droplets compared to microbubbles is that they can be made very small scale so eLiposomes can extravasate via the EPR effect.

(194g) Immunostimulating Effect of Ginseng Polysaccharide Nanoparticles On Macrophage Cell Lines

Akhter, K. F., Western University
Charpentier, P. A., Western University
Lui, E., Western University
Originally presented on: 11/4/2013 17:15:00 - 17:35:00

Ginseng polysaccharides, one of the main components of ginseng, provide unique biological activities such as immunostimulating effects. This work focuses on nanoprocessing of ginseng polysaccharides for enhancing the efficacy of active nutraceutical ingredients (API), and encapsulation within a carrier for controlled API delivery. Nanoparticles of ginseng polysaccharide extracts were prepared using a microfluidic device and compared to other conventional wet chemical processes such as anti-solvent and reverse microemulsion techniques. The morphology and size of the nanoparticles were characterized by SEM, TEM, FTIR, and DLS. Depending on experimental conditions, microfluidics was found to provide unimodal polysaccharide spheres down to 20 nm with very narrow particle size distributions. Nanosized ginseng polysaccharide extracts were then encapsulated with gelatin, to give 180 nm spheres suitable for integration into capsules for subsequent oral drug delivery. For bioassay and animal model studies, a simple and efficient procedure was developed for the fluorescent labeling of ginseng polysaccharides. This procedure utilized a one-pot strategy with fluorescein-5-thiosemicarbazide (FTSC) to introduce a thiosemicarbazide group onto the aldehyde group at the reducing saccharide end to form a stable amino derivatives via reductive amination. The polysaccharide-FTSC derivative was characterized by photoluminescence and fluorescence microscopy. These techniques showed that the dye was highly fluorescent, maintaining the nanostructure. These labeled ginseng polysaccharides are presently being examined for bioassay studies to evaluate biological activities (immunomodulation) of the nanosized polysaccharides in murine macrophage cell lines and animal studies using rats. This work shows that nanoprocessing of ginseng polysaccharides can enhance controlled nutraceutical release for immunomodulation.

(230a) Chromatographic Scale-Up On a Volume Basis

Hansen, E. B., Novo Nordisk

Originally presented on: 11/5/2013 8:30:00 - 8:55:00

The scale-up of a chromatographic separation is in principle a simple procedure since the process parameters are scalable in a linear fashion. The process is scaled by increasing the column diameter while keeping the bed height, the velocity, and the volumes of the different phases (measured in column volumes) constant. The problem with this simple and well established concept, usually referred to as linear scale-up, is the lack of flexibility.

For practioners, especially those dealing with large scale operations, it is no secret that chromatography columns come in discrete sizes regarding the diameter. The diameter and the number of columns are the only design variables we have available when going into production scale. As a consequence we most often cannot choose the column volume that matches the desired capacity. We are faced with the problem that for a chosen diameter we need for instance 2.2 columns in order to meet the capacity demands. The solution is either to invest in a third column or go back to the lab and try to optimize the process in order to squeeze it into two columns. Both solutions can be rather costly either in investment or in time consumption in development. Further, we have no guarantee that the optimization will actually succeed (not to speak of the documentation burden if it does). Even when we manage to meet the right capacity we still risk encountering the same problem at a later date if we want to increase the production capacity. In pilot scale the problem can be even more pronounced since we must use existing equipment. The lack of flexibility of the column volume can therefore increase the number of cycles on each column and thereby the duration of the pilot campaign (where the chromatography steps are often the bottleneck). At the other end of the scale, in the development phase, we are met with similar problems. Knowing that the process will end up in production some day and the bed height must be kept constant we aim for a suitable bed height (like 20 cm) when developing the chromatography in lab scale. In order to avoid packing problems and flow irregularities a certain column diameter (~ 1 cm) is required and we therefore end up with a lower volume limit around perhaps 15-20 mL in order to have a representative process step. Especially in the early phase of development where the product quantities are limited this can be a substantial volume which reduces the number of trials and we therefore risk a delay in development. A volume in this range is also certainly not desirable for special tasks like virus validation.

Fifteen years ago we revisited the scaling principles of chromatography in order to find a more flexible way of scaling up (or down). From a conceptual point of view the solution was quite simple and does not involve any new theory. By scaling the process based on a constant contact time (residence time) we can, within limits, introduce the column bed height as a process design variable rather than a fixed parameter. Since the bed height is a continuous variable, as opposed to the diameter, we can use it to obtain exactly the volume we need in any scale. After proof of concept in lab scale and later in pilot scale the new concept became an established approach which has proven extremely valuable over the entire range from development in lab scale to commercial production. We have previously presented and published some of our early results. Our expectation at the time was that this would become an established approach at least for pharma companies producing significant volumes. But our experience from interactions with companies, including producers of chromatography equipment, and with the regulatory authorities has shown that this is apparently still new territory. Since we find the approach very useful to the field of industrial chromatography we would like to take the opportunity to present ten years of experience in chromatographic scale up on a volume basis. The presentation will include the theoretical background and proof of concept as well as industrial cases with scaling factors of more than 1000. Examples include IEX, HIC, RPC, and SEC processes for the separation of both peptides and large proteins.

(230b) Measurement, Prediction, and Control of the Physical Stability of a Hygroscopic API

Boehmcke, P., Bristol-Myers Squibb
Domagalski, N., Bristol-Myers Squibb Company
Tabora, J. E., Bristol-Myers Squibb
Originally presented on: 11/5/2013 8:55:00 - 9:20:00

The selection of hygroscopic crystalline forms is sometimes necessary to enable the clinical development of API. In this work the drug substance isolated from the manufacturing process is isolated as an anhydrous crystalline form which readily converts to a hydrate under environments of moderate relative humidity. Form control during storage and drug product manufacture was necessary to ensure process robustness. This presentation will focus upon experimental and modeling studies to enable definition of a control strategy which preserves the anhydrous form of the API. This strategy was applied to packaging and storage of the drug substance as well as processing and sampling of the drug product.

The talk will discuss a experimental set up designed to measure the dynamic hydration under different relative humidity environments by automatically recording the relative humidity, temperature, and sample mass. Different geometries were used to examine the impact of cake length on hydration rate.

Finally, a model incorporating solid state reaction kinetics and powder vapor diffusion was developed from the experimental data. The model allowed for a quantitative assessment of the risk and impact of hydration under different manufacturing and storage environments.

(230c) The Application of Advanced Control and Model Based Strategies to Biopharmaceutical and Pharmaceutical Processes

Barrett, M., University College Dublin
Whelan, J., University College Dublin
Carr, M., APC Ltd
Glennon, B., UCD
Originally presented on: 11/5/2013 9:20:00 - 9:45:00

Pressures are increasing to develop, optimise and scale-up processes in minimal time frames whilst maximizing manufacturing related process understanding (QbD). In turn strategies such as process control and modelling are emerging as powerful platforms to compliment and maximise the output from current experimental programs.

Presented are novel model based control strategies that enable the streamlined development, scale-up and scale-down of bio/pharmaceutical unit operations. Cell culture and crystallization examples will be shown where the interrogation of these process development platforms has facilitated a fundamental understanding of the systems kinetic trajectory and environmental sensitivities. This process insight enabling better process titre, reliability, scale-up (cell culture) and particle shape, size and control (crystallization) to be delivered across development and manufacturing scales.

These platforms importantly combine both scale (i.e. mixing) and kinetic (i.e. secondary nucleation, decreasing product titre) related effects to facilitate the development of manufacturing based strategies within a laboratory environment. The introduction of 'virtual lab and plant' aspects of these platforms will also be presented whereby the predictive nature of these tools aids process design for scale-up / scale-down and tech transfer.

(230d) Modeling of the Secondary Drying and the Fate of Organic Solvents for Spray Dried Dispersion (SDD) Product

Hsieh, D., Bristol-Myers Squibb Company
Yue, H.,
Nicholson, S.,
Roberts, D., Bristol-Myers Squibb Co.
Schild, R., Bristol-Myers Squibb
Gamble, J., Bristol-Myers Squibb
Lindrud, M., Bristol-Myers Squibb Co.
Originally presented on: 11/5/2013 9:45:00 - 10:10:00

The spray dryer has been used to generate amorphous dispersions of API and polymer to enhance the solubility and bioavailability of drugs that are less water soluble. This powder, after spray drying, contains residual solvents and they need to be further removed to an acceptable limit in a secondary dryer. The purpose of this study is to understand the secondary drying mechanism and establish a model to predict the drying performance for scale-up and the fate of organic solvents.

The kinetics of solvent removal from Spray Dried Dispersion (SDD) product follows the Fickian diffusion behavior. Since the SDD particles are typically hollow and the shell thickness of the hollow bead is typically between 10-15% of the diameter of the bead, the effect of curvature can be neglected. Hence, a diffusion equation for the geometry of a slab as an approximation was used for modeling the secondary drying. Excellent agreement is obtained between the experimental data and the prediction from the modeling. The control parameters of the secondary drying are the diffusion coefficient and the diffusion length. Since the diffusion coefficient is a function of temperature only, the secondary drying performance is not scale dependent. This secondary drying has been successfully scaled up from a lab scale of a few grams to a bench scale of 500 grams in an agitated conical dryer and the large-scale drying kinetics follow the model prediction well. The fate of residual solvents including acetone, methanol and others can be successfully determined from this modeling work.

(230e) Complex Fitness Diagrams and Downstream Processing

Lightfoot, E. N., University of Wisconsin--Madison

Originally presented on: 11/5/2013 10:10:00 - 10:35:00

Focus here is development of production scale processes and specifically with modifying optimized laboratory processes that are unsuitable for large-scale operations. Our example is production of high-value fermentation produced biologicals such as therapeutic proteins, and we approximate value in terms of evolutionary fitness dagrams. We also suggest that speed and economy of development can often be ncreased quite significantly by judicious incorporation of heuristics. We show that one is rapidly led bythis kind of thinking to radical but highly beneficial new processes ffor industrial scale operation. Success un such a procedure rquires careful specification of system and goals.

(230f) Bilayer Tablet Compaction: Translation From a Single Stage Compaction Simulator to a Rotary Press

Rowe, J. M., Bristol-Myers Squibb
LaMarche, K., Bristol-Myers Squibb Company
Carroll, B., ExxonMobil
Pandey, P., Bristol-Myers Squibb
Subramanian, G. A., Bristol-Myers Squibb
Nikfar, F., Bristol-Myers Squibb
Originally presented on: 11/5/2013 10:35:00 - 11:00:00

Fixed dose combination products (FDC) are formulations containing two or more active pharmaceutical ingredients (API) combined within a unit dose. In recent years, FDCs have become increasingly popular in several therapeutic areas because they offer convenience, a reduced pill burden, and cost savings for patients.

In cases when the APIs are chemically incompatible or multiple release profiles are desired, the dose must be formulated as a layered tablet. However, layered tablets pose unique development and manufacturing risks because of the fine balance required between maintaining sufficient physical separation between the layers to mitigate chemical incompatibilities and ensuring that the tablet remains intact (i.e. avoid layer delamination) throughout the manufacturing process and product shelf life. Challenges during the manufacture of layered tablets can be difficult to resolve due to the limited mechanistic understanding of the compaction process. Thus, the development process for layered tablet formulations has been largely trial-and-error, requiring significant resource investments, especially when considering that layered tablet production on a rotary press requires kilogram-quantities of materials.

In this work, the authors have sought to gain mechanistic understanding of the compaction process for layered tablets to minimize resources required during product development, more specifically to better understand how experimental factors (i.e. material selection, compaction conditions) impact layer interface strength. Since the failure of layered tablets typically occurs due to delamination at the interface, a fundamental appreciation of the governing mechanisms that impact interface strength is crucial in order to improve the overall mechanical integrity of that layered tablet. A single station compaction simulator, fitted with a three-sensor instrumented die, was used as a material-sparing tool (requires only grams of material) for the development of bilayer tablets and to increase fundamental understanding of the compaction process. The instrumented die was used to measure the die-wall pressure exerted by each layer of the bilayer tablet during the compaction process.

Bilayer combinations of placebo and active granulations over a range of compaction forces were examined in this study. Young's moduli and Poisson's ratios, as well as additional elastic material properties described by the Drucker-Prager-Cap model, were determined using measured die-wall pressure profiles for each material in the study. Analysis of the die-wall pressure profiles during bilayer compaction yielded a correlation between the residual stress within each of the layers after compaction and the interface strength of the bilayer tablet. The level of residual stress experienced by different materials, when compressed as either the first or second layer in a bilayer tablet, was related back to elastic material properties. Using this knowledge, select material combinations were compressed as bilayer tablets on a rotary tablet press. The trends in tablet interface strength for tablets produced on the compaction simulator were similar for those manufactured on the rotary press. However, it was

noted that for certain material combinations, tablets made on the compaction simulator and the rotary press did

not possess the same solid fraction when compacted under the same forces. When tablet strength was examined

as a function of solid fraction instead of compaction force, the compaction simulator was found to adequately

emulate the rotary press. Discrepancies between press compaction force and corresponding solid fractions were

related back to elastic material properties. In conclusion, the compaction simulator was found to be a useful

tool to develop bilayer tablet formulations at small scales, providing results that may be translated to larger scale

production on a rotary press.

(260a) Applications of Modeling and Simulating Moisture Profiles During Fluid Bed Granulation Process Development

Mehrman, S., PDMS
Cunningham, J., PDMS
Sanchez, E., Janssen Pharmaseuticals
Futran, M., Bristol-Meyers Squibb
Murphy, T., PDMS
Weber, J., PDMS
Fermier, A., Janssen Pharmaceuticals
Boucher, R.,
Skalkin, A.,
Originally presented on: 11/5/2013 8:30:00 - 8:55:00

Developing a robust fluid bed granulation process is often challenging due to a variety of equipment configurations and interrelated controls. To overcome these challenges we have been developing first principal modeling and simulation tools focused on moisture control profiling. Examples using these tools to model historical batch performance and simulation to design scale up and process DOE's will be described.

(260b) Understanding Micro Environmental Thermodynamics of Tablet Coating Process Using Data Logger - Case Studies

Subramanian, G. A., Bristol-Myers Squibb
Pandey, P., Bristol-Myers Squibb
Ji, J.,
Gour, S., Bristol-Myers Squibb
Zacour, B., Bristol Myers Squibb
Originally presented on: 11/5/2013 8:55:00 - 9:20:00

Pharmaceutical film coating process is typically monitored by inlet and exhaust temperature and humidity conditions, with measurement devices located outside the coating pan. Current configurations of coating pans does not allow continuous measurement of the temperature and humidity within the tablet bed. Measurable differences in heat and mass transfer efficiencies across coating pan sizes and designs can result in different drying and film forming dynamics at the same measured outlet temperature and humidity. Similarly, different microenvironment (temperature and humidity conditions inside the tablet bed) and hence film forming dynamics can be expected for coating operations carried out at the same exhaust temperature, if this temperature is achieved by changes to the process parameters such as inlet air volume, air temperature, suspension spray rate or suspension solids concentration. Considerable insight into the dynamics of the coating process can be gained if data loggers can be incorporated in the tablet bed and placed such that they can collect temperature and humidity conditions in the tablet bed continuously. Two case studies will be presented where the micro-environmental data from PyroButtons provide meaningful correlation to coated tablet defects. The authors will show that this data can be used to establish a process design space for a given coating and tablet formulation. The first case study is a full factorial design of experiments (DOE) on three process parameters; exhaust temperature, pan speed and spray rate conducted on a 24 inch pan coater, that showed correlation of micro environmental conditions to coated tablet appearance. The second case study will show correlation between delamination tendency of a bilayer tablet and tablet-bed micro environment.

(260c) An Integrated Model for the Thermodynamics of the Film Coating Operation the Mixing Phenomena in the Drum and the Distribution of Properties in the Coated Tablet

Garcia-Munoz, S., Pfizer Worldwide R&D
Cano, A., Process Systems Enterprise, Inc.
Li, J.,
Blackwood, D. O., Pfizer Worldwide Research and Development
Berchielli, A.,
Originally presented on: 11/5/2013 9:20:00 - 9:45:00

The film coating operation is a versatile step in the manufacture of solid dosage forms of drug product in the pharmaceutical industry. The coat applied to a drug product can be required for either pure elegance purposes (non-functional coat), or to determine critical functionality of controlled release dosage form; it may also be used to coat the drug itself onto a core carrier. The distribution of the coating material for a functional coating is a critical quality parameter due to the direct link with the release of drug from the tablet. For non-functional coatings, the distribution (and level) of condensable components into the tablet core can play a critical role in the long term physical and chemical stability of the product.

In a previous work we presented a dynamic model of the film coating system that aimed at considering all the transient energy and mass transport phenomena that occurs during the execution of a given recipe for this process. This contribution describes the modeling of the mixing phenomena inside the coater. The mixing model was validated with data acquired with a digital video system and tracer tablets for 20 experiments for coaters from laboratory scale to large-scale commercial manufacture. The parameters of the mixing model are in line with prior knowledge about the mixing behavior of tablet beds. Furthermore, the mixing model was coupled with the dynamic thermodynamic model to provide a holistic modeling approach to the film coating operation where the distribution of the properties of the coated tablets (e.g. coating content, water absorbed and coating thickness) can be predicted as a function of the operating recipe.

(260d) Experimental and Theoretical Study of Tablet Flow in a Lab Scale Pan Coater

Toschkoff, G., RCPE GmbH

Böhling, P.,

Johnson, D., Merck & Co., Inc.

Sun, J., Merck & Co., Inc.

Silverman, A. R., Merck & Co., Inc.

Scharrer, G., RCPE GmbH

Khinast, J. G., Graz University of Technology

Rajniak, P., Merck & Co., Inc.

Originally presented on: 11/5/2013 9:45:00 - 10:10:00

The coating of tablets using a film coating operation has become a common process in the pharmaceutical industry to add color distinction, physical protection, or taste masking properties to the final product. Given the delicate nature of some tablets, it is important to know how certain process parameters may affect the tablet motion and interactions within a coating pan. In recent years, *in silico* computer simulations have become a common trend for trying to predict this behavior using techniques like Discrete Element Methods (DEM) to compare/calibrate to actual experimental/physical conditions.

The aim of this work is to develop DEM simulations for a laboratory film coating pan (Labcoat 24? O'Hara Technologies, USA) and calibrate the model using experimental measurements from the same instrument. The model was developed around a 12-16 kg fill of the coating pan with 10/32? Round Standard Curve placebo tablets, with comparisons being performed on the distribution of spray zone residence time of selected tablets and the bed cycle time.

Experimental data was generated using tracking tracers and a video imaging system that was placed inside the coater. The images used a rectangular region of the tablet bed surface as the spray zone, and the frequency of the tracer in the spray zone was extracted using automated image processing, see Fig.1.

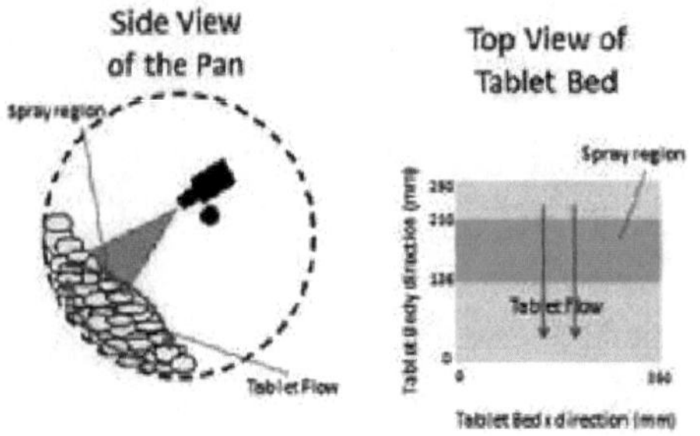

Fig. 1: Schematic of the experimental setup; the same size and location of the spray zone is also used in the DEM simulations.

Simulations were performed using the commercial software EDEM 2.4 (DEM Solutions, UK) where the tablet shape was approximately reproduced using 8 partially intersecting spheres (?glued sphere? approach). A post-processing algorithm was developed featuring the same zone as was imaged as the spray zone in the experiments to track the spray and bed residence times of selected tablets.

The simulation results showed good agreement with the experimental data, helping to calibrate and validate the model and its parameters. The validated DEM simulation can be used to show local tablet velocity and collision forces, and how they relate to the operating conditions of the coating pan. With further development of this DEM approach and understanding the hardness of different tablet systems, it may become possible to design and optimize coating pans and process conditions for ideal coating operations with minimal tablet damage.

(260e) Use of Mechanistic Modeling to Understand Tablet Die Filling

Smith, R. B.,
Dubey, A., Tridiagonal Solutions Inc.
Muzzio, F. J., Rutgers, The State University of New Jersey
Vedapuri, D., Tridiagonal Solutions Inc.
Originally presented on: 11/5/2013 10:10:00 - 10:35:00

Tableting of granular material is ubiquitous in a wide range of industries – pharmaceutical, consumer goods, food, specialty chemicals, powder metallurgy and coal energy. Advancement in computing have allowed for high fidelity modeling and simulation of the die filling to study the impact of key processing variables on the tableted product. A combination of Discrete Element Method (DEM) and Computational Fluid Dynamics (CFD) has been used to gain valuable insights into the process. Prior work in this area has used simpler 2D models. This study employs detailed 3D models to study the effect of particle shape, size, size distribution, die-fill speed (relates directly to scale up), cohesiveness of the material, and the filling method (gravity vs. suction fill). A smaller particle size results in a higher solid fill fraction in mono-dispersed particle sizes. A wide particle size distribution leads to higher solid fill as the fines enter the interstitial spaces, when compared to mono-dispersed particles. Irregular shaped particles, a reality of many processes, lead to lower solids fraction. Higher than optimal die filling speeds lead to inefficient filling which is also a function of the material flow properties, while using a suction fill greatly enhances the solids fill fraction. The trends obtained by these models are compared with the available experimental data. The study also compares filling with and without the use of air (CFD) and the effect of using a one-way vs. a two-way DEM-CFD coupling. This study has a potential of predicting properties and issues like stratification, solubility, content uniformity, mechanical strength of tablets, and can be extended to study more advanced applications such as bi-layer tablets.

(260f) Application of Fracture Mechanics Model for the Characterization of Bilayer Tablets

Kottala, N., Bristol-Myers Squibb Comany
Abebe, A.,
Sprockel, O. L., Bristol-Myers Squibb
Akseli, I., Boehringer Ingelheim
Nikfar, F., Bristol-Myers Squibb
Cuitino, A., Rutgers University
Originally presented on: 11/5/2013 10:35:00 - 11:00:00

Application of Fracture Mechanics Model for the Characterization of Bilayer Tablets

Bilayer tablets are generating great interest recently in the pharmaceutical industry as they offer several advantages over conventional single layer tablets. However, the production of bilayer tablets has been facing great challenges as the layered tablets are prone to delaminate at the interface and fracture in the individual layers due to insufficient bonding strength. In the pharmaceutical industry, the process of bilayer design has been heavily dependent on the trial-and-error approach during the formulation and process development stages. To overcome this hurdle, it is critical to estimate and understand the impact of mechanical properties (bulk modulus and Poisson's ratio) on the performance of bilayer tablets and to develop a predictive model for the characterization of bilayer tablets.

This presentation will focus on the methodology and the model that has been developed based on the principles of fracture mechanics for the characterization of bilayer tablets. As part of this endeavor, interfacial stress intensity factor was determined. The estimated stress intensity factor using the fracture mechanics model will provide insight into the interplay of material properties, bilayer interfacial topography, and processing parameters on the performance of bilayer tablets. The mechanistic understanding and the methodology developed will enable to move away from the existing "trial-and-error" approach during the design and development of bilayer tablets. The new paradigm of bilayer tablet development will incorporate the principles of the quality by design by leveraging the *a priori* knowledge.

(316a) Formation of Poly(-caprolactone) Nanoparticles in Microstructured Mixers Via Solvent Replacement Precipitation: Fundamental Studies and Scale-Up Concepts

Kolbl, A., Karlsruhe Institute of Technology
Qayyum, A.,
Kraut, M., Forschungszentrum Karlsruhe GmbH
Dittmeyer, R., Karlsruhe Institute of Technology
Originally presented on: 11/5/2013 12:30:00 - 12:55:00

Poly(ε-caprolactone) nanoparticles are frequently employed in pharmaceutics for controlled drug release [1] and are often obtained via solvent replacement precipitation [2]. With microstructured mixers these particles were obtained free from mixing influences during their formation (size app. 100 - 500 nm). This is shown by two different approaches: 1.) the variation of the feed rates shows no influence on the particle size distribution. 2.) precipitation with different mixers (cyclone mixers [3] and multilamination mixers [4]), which could be clearly distinguished regarding their 'mixing qualities' using the iodide iodate reaction method, yield precipitates with the same particle size distribution. Hence, mixing in these microstructured mixers is significantly faster than the particle formation rate.

Fundamental studies:

Ruling out inhomogeneities caused by insufficient mixing, influencing factors on the particle size were examined using a microstructured cyclone mixer. With this mixer excellent mixing is obtained at low flow rates. This suggests that microstructured cyclone mixers are an effective tool for scientific research, especially when materials are expensive or difficult to obtain. Influencing factors which were identified in previous studies [2] are substantiated, namely the polymer outlet concentration and the composition of the solvent (acetone and water in different ratios). Examinations free from mixing influence allow for the quantification of these factors. It could be shown that the reproducibility increased inversely to particle size.

Scale-up concepts:

Mixing in multilamination mixers (V type and Y type) is significantly slower compared to mixing in cyclone type mixers but still significantly faster than the particle formation rate. IMVT's multilamination mixers are constructed from stacks of micromachined foils. By stacking micromachined foils a multichannel system is obtained. Scaling of those mixers is possible by the variation of the number of micromachined foils. Constant 'mixing qualities are obtained keeping a constant linear fluid velocity in the microchannels. The scaling concept has been investigated using the iodide iodate reaction method [4].

[1] R. Langer, Science 249 (1990) 1527-1533.

[2] F. Lince et al., Journal of Colloid and Interface Science 322 (2008) 505–515.

[3] A. Kölbl et al., Chemical Engineering Journal 167 (2011) 444–454.

[4] A. Kölbl et al., Chemical Engineering Journal 160 (2010) 865–872

(316b) Bioreactor Simulation With Cuda

Witz, C.,
Tantikul, T., Graz University of Technology
Khinast, J. G., Graz University of Technology
Originally presented on: 11/5/2013 12:55:00 - 13:20:00

Bioreactor simulation with CUDA

Christian Witz[1], Tawan Tantikul[1], Johannes Khinast[1,2]

[1]Graz University of Technology, Institute for Process and Particle Engineering

[2]Research Center Pharmaceutical Engineering

Abstract

The effectiveness of a biopharmaceutical manufacturing process depends to a large extent on the efficiency of the bioreactor, especially in the field of generic drugs. So far, the engineering process of the reactor design has been mostly driven by empirical knowledge, as the simulation of this complex multiphase and multiscale process was impossible for many years. However, despite the recent improvement of computational capabilities, the simulation of an industrial scale reactor takes months for only a few seconds of real operation time.

Thus, the goal of this study is to use graphic cards to speed up this simulation. The Compute Unified Device Architecture (CUDA) technology of nVidia has made the computational power of graphic processing units (GPUs) available for scientific calculations [1]. In the multiphase simulation, the large number of computing units in the GPU leads to a significant reduction of calculation time. To archive this acceleration, an efficiently parallelizable simulation method is needed.

The lattice Boltzmann method (LBM) which was developed based on the lattice gas automata [2], represents an efficient way to numerically capture the liquid phase flow dynamics on the GPU computing machine. It uses a regular grid with evenly distributed nodes. To model the geometry without changing the regular grid the immersed boundary method is used. The air bubbles are simulated with the Lagrangian particle tracking (LPT) method. The sum of the forces acting on each bubble, i.e. the drag, the buoyancy, the lift force, the history force, the added mass effect and gravity is used to determine the acceleration of the particle. The acceleration and the time step length give the velocity and the position change at the end of the time step.

For coalescence and breakup stochastic models are used. The phases are coupled with a two way approach.

As the calculation results are computed on the graphic card, simulation data can be visualized real time during the calculation. Furthermore to simulate large reactors, the code has a multi GPU functionality, hence, it can distribute the workload of the simulation on several graphic processors. This multi GPU code is able to handle the large scale, memory demanding simulation such as the simulation of the industrial scale reactor.

References

[1] http://www.nvidia.com/cuda

[2] Sukop, M.: Lattice Boltzmann Modeling, Springer, Berlin, 2007.

Submission details

Session - 26013 Mixing Issues and Scale-Up/Scale-Down Strategies for the Production of Pharmaceutical and Biopharmaceuticals

Emails:

christian.witz@tugraz.at
tawan.tantikul@tugraz.at
khinast@tugraz.at

(316c) Characterization of Kla Using Definitive Screening Designs for Fermentation Development and Scale-Up

Ornek, D.,
Anderson, R.,
Chopra, M.,
Lin, W.,
Ramsey, P.,
Originally presented on: 11/5/2013 13:20:00 - 13:45:00

Characterization of $k_L a$ using Definitive Screening Designs for Fermentation Development and Scale-Up

Fermentation development and optimization are typically performed at small scale to reduce their cost and time requirement. High cell density aerobic fermentation is scaled-up often based on keeping volumetric mass transfer coefficient ($k_L a$) constant. To facilitate a successful transfer of an optimized small-scale fermentation process to lab, pilot, and industrial scale, the $k_L a$ profiles for these fermentors should be determined. Although a number of correlations and predictive models for $k_L a$ are available, they do not have universal applicability.

The purpose of this study was to use a statistical Design of Experiments (DoE) approach to characterize $k_L a$ in 1.3L DASGIP, 10L Biostat B, 40L Biostat CDCU and 200L ABEC microbial fermentors, and hence to use this information for fermentation scale-up. Polarographic dissolved oxygen sensors were used to measure dissolved oxygen concentrations in the fermentors. $k_L a$ was determined in de-ionized water by using the dynamic re-oxygenation method. Definitive Screening Designs, a new efficient statistical Design of Experiment, were used to characterize $k_L a$ in the fermentors. Four experimental factors, temperature, agitation, air flow rate, and volume, were included in the experimental design as input parameters, and $k_L a$ was the output parameter. Definitive Screening Designs showed that agitation, air rate and volume were statistically significant for all the fermentors. Air and volume had strong positive correlations with $k_L a$, and agitation had strong quadratic effects on $k_L a$. Results showed that four fermentors have comparable $k_L a$ profiles. The correlations determined here will be used as an effective tool for fermentation development, scale up and tech transfer.

(316d) Scaling Down the Production of Biopharmaceuticals to Continuous Micro-Chips

Alvarez, M. M., Tecnologico de Monterrey
Carrillo-Cocom, L. M., Tecnologico de Monterrey
Garza-Garcia, L. D., Tecnologico de Monterrey
Tapia-Mejia, E. J., Tecnologico de Monterrey
Camacho-Leon, S., Tecnologico de Monterrey
Rodriguez-Gonzalez, C. A., Tecnologico de Monterrey
Garcia-Lopez, E., Tecnologico de Monterrey
Originally presented on: 11/5/2013 13:45:00 - 14:10:00

We report the use of continuous flow micro-devices for the production of a glycosylated pharmaceutical product (a monoclonal antibody). We culture CHO cells on the surface of different PMMA/PDMS micro-channel chips. In particular, three different channel configurations were analyzed: (a) an ondular channel design, (b) a "zigzag" shaped micro-channel, and a (c) series of shallow compartments connected with a chanel. The surfaces of all micro-channels were textured with semi-spherical cavities and coated with fibronectin. Experiments of continuous perfusion at different flow rates were performed in the range of 0.1 to 6.0 microL/min. The velocity field at the different flow rates and geometries was investigated through CFD. Significant differences in attached cell density, cell proliferation, and monoclonal antibody production were observed among the different experimental scenarios. In some configurations, we observed sustained monoclonal antibody productions resulting in a steady-state productivity above 6,240 mg L^{-1} day^{-1}, at least 100 times higher than that observed in commercial scale fed-batch systems. We discuss the effect of surface texture, system geometry and flow rate on the performance of these micro-bio-reactors.

(316e) Computing Dissolved Oxygen Profiles in Aerobic Fermenters

Kasat, G., Tridiagonal Solutions Inc.
Vedapuri, D., Tridiagonal Solutions Inc.
Benz, G., Benz Technology International, Inc.
Originally presented on: 11/5/2013 14:10:00 - 14:35:00

Aerobic fermenters have been a staple method of bioprocessing for many years. Oxygen is necessary for all aerobic fermentation. Maintaining the appropriate and uniform concentration of dissolved oxygen (DO) is important for the successful growth of culture. Large DO gradients can lead to operating problems. Low or High DO levels in some part of reactors may lead to bi-product formation. A priori information on DO distribution can help in proper design and scale-up of industrial fermentation process.

Until now, there has been no way to reliably predict DO distribution. Indeed, it has not been possible to even measure it. A typical aerobic fermenter has only one DO probe; a few rare fermenters have two DO probes: one near the top of the fermenter, and one near the bottom. A substantial DO gradient has been observed in some fermenters: up to 100:1 ratio of the bottom DO to top DO with multiple radial turbines and up to 2:1 with axial/radial combinations.

Computational Fluid Dynamics (CFD) models have been successfully used to predict mixing, gas hold-up distribution, mass transfer & heat transfer rates in stirred gas-liquid fermenters. In this study the CFD models were used to predict DO distribution in a fermenter. Eulerian-Eulerian multiphase model along with standard k-e turbulence model has been used to predict the turbulent gas liquid flow in the vessel. The results of the CFD model were used to calculate oxygen transfer rates and DO distribution in the vessel. While calculating DO it has been assumed that the oxygen consumption rate (OUR) is uniform throughout the vessel.

Multiple simulations were carried to understand the impact of agitator design and operating condition on the DO distribution. Since the actual DO distribution cannot be experimentally verified, the absolute magnitudes of the results may not be accurate. However, it can be said that CFD models can be successfully used to predict DO distribution profiles in aerobic fermenters.

(316f) Hydrodynamic Effects of An Arch-Shaped Fiber Optic Probe in a USP Dissolution Testing Apparatus 2

Zhang, Y., New Jersey Institute of Technology
Bredael, G., Merck & Co., Inc.
Armenante, P. M., New Jersey Institute of Technology
Originally presented on: 11/5/2013 14:35:00 - 15:00:00

Dissolution testing is widely used in the pharmaceutical industry to evaluate newly developed drug formulations and as a quality control method to insure that solid dosage forms have consistent dissolution property. Typically, samples are manually drawn from the dissolution vessel prior to analysis. An approach to overcome the limitations of manual sampling consists in the use of sampling probes, such as fiber optic probes, permanently inserted in the dissolution medium and continually sampling the drug concentration in it as the solid dosage form dissolves. Despite their advantages, permanently inserted fiber optic probes can alter the normal fluid flow within the vessel and produce different dissolution testing results.

In this study, the hydrodynamic effects introduced by an arch-shaped fiber optic probe in a USP Dissolution Testing Apparatus 2 are studied by: (1) conducting dissolution tests, with and without the probe, using Prednisone tablets fixed at nine different locations at the bottom of the vessel and comparing the dissolution profiles obtained using statistical tools; and (2) experimentally determining the velocity profiles in the vessel, with and without the probe, using Particle Image Velocimetry (PIV) and quantifying changes in the flow velocities on selected horizontal iso-surfaces.

The results show that the arch shaped fiber optic probe does have a baffling effect on the hydrodynamics in the dissolution vessel. This effect results in changes in the velocities in the fluid flow, and therefore in changes in the dissolution rate of the tablets undergoing testing. The baffle effect is observed mainly in the region where the probe is inserted. However, this perturbation is also found to reach the region below the impeller and to change the velocity profile there, resulting in differences in dissolution profiles when the tablets are fixed at positions that are downstream of the probe and within the low velocity region below the impeller.

On the other hand, the hydrodynamic effect generated by the probe does not appear to be particularly strong. In most dissolution testing runs, the changes in dissolution profile are not large enough to fail the tests, according to the FDA criteria (f_1 and f_2 values). The PIV measurements additionally show that the baffle effect is not strong enough to break the overall flow pattern, or to affect the region around the impeller, which is dominated by the main flow generated by the impeller.

It can be concluded that the hydrodynamic effects generated by the arch-shaped fiber optic probe are real and observable, resulting in slightly modification of the fluid flow in the dissolution vessel and therefore in detectable differences in the dissolution profiles. However, these effects are limited and do not typically lead to dissolution testing failures.

(317a) Modelling the Manufacturing Process and Product Performance of Roller Compacted Pharmaceutical Tablets

Gavi, E.,
Reynolds, G. K., AstraZeneca
Pinto, M. A., Process Systems Enterprise
Bermingham, S. K., Process Systems Enterprise
Originally presented on: 11/5/2013 12:30:00 - 12:55:00

Direct compaction, dry granulation and wet granulation are the three processes typically employed to produce pharmaceutical tablets. Dry granulation is used to improve content uniformity, dissolution times and stability, by aggregating particles in granules without water addition. This is done by compressing the powders (active ingredient, excipients, etc.), previously mixed, in a roller compactor. The ribbon produced is then milled to the required granule size before being compressed in a tablet press.

In this presentation, a flowsheet of a tablet manufacturing process via dry granulation is developed. Literature models for roller compaction (Johanson, 1965; Reynolds et al, 2010) and milling (Vogel and Peukert, 2005) are used. A new model for tablet compaction is developed to predict tablet porosity and tensile strength as a function of tablet compaction pressure. This model is based on concepts presented in the literature previously (Wu et al, 2006; Farber et al, 2008). In order to evaluate the influence of material properties and process operating conditions on the produced tablet, a model for tablet dissolution is also developed. This model extends work done previously (Wilson et al, 2012), and it relates the tablet porosity and material properties to the tablet erosion rate and dissolution time.

The individual unit operations in the process are validated against experimental data. The validated process flowsheet is used to perform sensitivity studies to explore the effect of roller compaction operating pressure and tablet compaction pressure on ribbon and tablet porosity, tablet tensile strength and dissolution time.

References

Johanson JR (1965) A rolling theory for granular solids. ASME, Journal of Applied Mechanics Series E 32, 842-848

Reynolds G, Ingale R, Roberts R, Kothari S, Gururajan B (2010) Practical application of roller compaction process modelling. Computers and Chemical Engineering 34, 1049?1057

Vogel L, Peukert W (2005) From single particle impact behaviour to modelling of impact mills. Chemical

Engineering Science 60, 5164 ? 5176

Wu C-Y, Best SM, Bentham AC, Hancock BC, Bonfield W (2006) Predicting the tensile strength of compacted multi-component mixtures of pharmaceutical powders. Pharmaceutical Research 23, 1898-1905

Farber L, Hapgood KP, Michaels JM, Fu X-Y, Meyer R, Johnson M-A, Li F (2008) Unified compaction curve model for tensile strength of tablets made by roller compaction and direct compression. International Journal of Pharmaceutics 346, 17?24

Wilson D, Wren S, Reynolds G (2012) Linking dissolution to disintegration in immediate release tablets using image analysis and a population balance modelling approach. Pharmaceutical Research 29, 198-208

(317b) Modeling Drug Precipitation in the Human Gastrointestinal Tract Using gCOAS

Abbou Oucherif, K., Purdue University
Lu, J.,
Bermingham, S. K., Process Systems Enterprise
Braido, D., Rutgers University
Calado, F., Process Systems Enterprise
Taylor, L., Purdue University
Litster, J. D., Purdue University
Originally presented on: 11/5/2013 12:55:00 - 13:20:00

Over the last decade there has been a significant increase in the number of poorly soluble drugs in the pharmaceutical industry.[1] Poor solubility, especially for weakly basic compounds, can lead to crystallization *in vivo*and consequently to poor bioavailability.

Upon oral administration, weak bases can experience major pH swings when transiting from the acidic environment of the stomach to the alkaline milieu of the small intestine. This can lead to crystallization and precipitation of the compound reducing its bioavailability and extent of absorption.[2] Several computational oral absorption models have been developed over the last decade to predict the extent of oral absorption *in vivo*. However, while these models have successfully incorporated most of the pharmacokinetic and pharmacodynamic processes taking place in the GI tract,[3]they lack a scientific rationale to properly model drug precipitation. The onset of precipitation is often dictated by a static precipitation time rather than a supersaturation driving force.

The present work aims to use a novel *in silico* computational oral absorption model, gCOAS (Process Systems Enterprise Ltd, London, UK), to simulate drug precipitation *in vivo* with a strong emphasis on first principles modeling of crystallization kinetics. gCOAS is inspired by the Sugano framework[4] and offers a flexible modeling platform to model the different processes taking place *in vivo*. A key feature of the software is the ability to customize the segmentation and physiological parameters of the GI tract. Each compartment in the model incorporates a population balance model, and kinetic expressions for transit, dissolution, nucleation, growth, and permeation. The model also includes mass and charge balances of both drug species (ionic, non-ionic and micellar) and physiological ions in solution in the GI tract. In addition, to properly model drug transit and fluid flow, the volume of fluids in each compartment is dynamically modeled and is considered to be dependent on the volume of fluid uptake and the volume of endogenous secretions such as saliva, pancreas, mucous, and bile secretions.

In this study, dipyridamole, a weakly basic BCS class II compound, is chosen as the model active pharmaceutical ingredient. Solubility and nucleation threshold experiments are conducted under varying pH and bile salt concentrations to develop a solubility-supersolubility phase diagram for dipyridamole to serve as a design space for the nucleation and growth experiments.

Kinetic expressions using the experimental nucleation and growth data are then integrated in gCOAS. The flexible platform of the latter allows for the implementation of custom nucleation and growth kinetic models to account for the effect of bile salts on the crystallization kinetics.

References

1. Hauss, D. J. Oral lipid-based formulations. *Advanced Drug Delivery Reviews* **59**, 667–76 (2007).

2. *Modern Pharmaceutics*. 864 (CRC Press: 2002).doi:10.1201/9780824744694

3. Agoram, B., Woltosz, W. S. & Bolger, M. B. Predicting the impact of physiological and biochemical processes on oral drug bioavailability. *Advanced Drug Delivery Reviews* **50**, S41–67 (2001).

4. Sugano, K. Introduction to computational oral absorption simulation. *Expert opinion on drug metabolism & toxicology* **5**, 259–293 (2009).

5. Yu, L. X., Crison, J. R. & Amidon, G. L. intemational journal of Compartmental transit and dispersion model analysis of small intestinal transit flow in humans. *International journal of pharmaceutics* **140**, 111–118 (1996).

(317c) A Molecular Design Approach for Drug Tablet Formulation and Impact On Personalized Medicine

Pavurala, N., Virginia Tech
Achenie, L., Virginia Polytechnic Institute and State University
Originally presented on: 11/5/2013 13:20:00 - 13:45:00

A molecular design approach for drug tablet formulation and impact on personalized medicine

Naresh Pavurala, Luke E.K Achenie

Department of Chemical Engineering, Virginia Polytechnic Institute and State University, Blacksburg, Virginia 24060

One of the major challenges faced by the pharmaceutical industry is to accelerate the drug innovation process and reduce the time-to-market for new drug developments. This involves billions of dollars of investment due to the large amount of experimentation and validation processes involved. A computational modeling approach, which could rapidly explore the design space, reduce uncertainty and make better, faster and safer decisions, fits into the overall goal and complements the drug development process.

In the drug discovery and development process, the tablet formulation (oral drug delivery) is an important step. The tablet consists of active pharmaceutical ingredient (API), excipients and polymer. A controlled release of drug from this tablet usually involves swelling of the polymer, forming a gel layer and diffusion of drug through the gel layer into the body. The polymer is mainly responsible for controlling the release rate (of the drug from the tablet), which would lead to a desired therapeutic effect on the body.

In this presentation we propose a molecular design strategy for generating molecular structures of polymer candidates with desired properties estimated through structure-property models. Our work is expected to aid in generating the most favorable polymers based on the biopharmaceutical and pharmacokinetic properties. This could help in precipitating development failures in the early stages of drug discovery as well as reduce the very expensive late-stage failures.

In greater detail, we utilized group contribution models to estimate several desired polymer properties such as grass transition temperature (T_g), density (ρ) and linear expansion coefficient (α). We subsequently solved an optimization model, which generated molecular structures of polymers with desired property values. Some of the polymer repeat-units are already available in the literature; these include polyvinyl alcohol (PVA) and polyvinyl chloride (PVC) repeat-units. Examples of new repeat units are ?*CH=CH-COO*?, ?*CHCl*?CH_2?*CHCl*?. These repeat-units could potentially lead to novel polymers with interesting characteristics; a polymer chemist could further investigate these. We recognize the need to develop group contribution models for other polymer properties such as porosity of the polymer and diffusion coefficients of water and drug in the polymer, which are not currently available in literature. We expect our modeling approach to aid in the drug development process and contribute to the development of personalized medicine.

(317d) An Optimization-Based Strategy to Handle Missing Data in Partial Least Squares

Harinath, E., Carnegie Mellon University
Garcia-Munoz, S., Pfizer Worldwide R&D
Biegler, L. T., Carnegie Mellon University
Originally presented on: 11/5/2013 13:45:00 - 14:10:00

An Optimization-based Strategy to Handle Missing Data in Partial Least Squares

E. Harinath, S. Garcia-Munoz and L. T. Biegler

Latent variable methods have proved to be a powerful approach for the development of data-driven process models. The objective of these methods is to find hidden structures of large data sets consisting of highly correlated variables. The basic idea of these techniques is to decompose regression space X, or both the regression space and response space Y, into subspaces which are spanned by base vectors, called scores. For an example, in Principal Component Regression (PCR) method, score vectors of X, called principal components, are found by maximizing variance in X space. Then the principal components are regressed on Y space to predict response. However, PCR will not use any information available in Y space to find dominant direction but it simply explains only X space. In contrast to PCR, Partial Least Square (PLS) methods find score vectors for both X and Y, by maximizing covariance of X and Y. Thus PLS finds the dominant direction common to both X and Y spaces.

The most popular method for finding the score and loading vectors in PLS is the Nonlinear Iterative Partial Least Squares (NIPALS) algorithm. The key step of NIPALS is the deflation process in each iteration of the algorithm. In [1], a version of NIPALS algorithm is presented and analyzed its basic properties such as orthogonality of score and weighting vectors for X space. These properties are mainly due to the deflation process in the algorithm.

In process industry, there are many situations where input output data sets are having missing data points. For example, when sensors fails or are taken for maintenance, and loss of historical data or missing experiments [2-3], input output data lead to incomplete data sets. With incomplete data sets, the NIPALS algorithm may fail to provide the orthogonality properties of scores and weightings, which are essential for better interpretation of the resultant model. In [3], it is shown for PCA with incomplete data sets that the loading and score vectors determined by a modified NIPALS algorithm are not orthogonal.

To address problems associated with the modified NIPALS for PCA, an efficient nonlinear optimization framework is presented in [3]. The deflation process in PLS is difficult to formulate in an optimization framework in case of incomplete data. In [4], different optimization frameworks are discussed for multivariate regression methods. In particular, the *undeflated*PLS (UDPLS) method is presented in order to remove the deflation process for complete data set problem. In this work, we develop an UDPLS technique for missing data problem. This work is in line with the work presented for PCA in [3]. The proposed technique is based on Nonlinear Programming (NLP) methods. We present simulation case studies where we compare the parameters obtained from both the modified NIPALS and proposed algorithms, and demonstrate the effectiveness of our optimization-based approach.

[1] Agnar Höskuldsson. PLS regression methods. Journal of Chemometrics, 2(3):211–228, 1988.

[2] Philip R.C. Nelson, Paul A. Taylor, and John F. MacGregor. Missing data methods in PCA and PLS: Score calculations with incomplete observations. Chemometrics and Intelligent Laboratory Systems, 35(1):45 – 65, 1996.

[3] Rodrigo López-Negrete de la Fuente, Salvador García-Munoz, and Lorenz T. Biegler. An efficient nonlinear programming strategy for PCA models with incomplete data sets. Journal of Chemometrics, 24(6):301–311, 2010.

[4] Alison J. Burnham, Roman Viveros, and John F. MacGregor. Frameworks for latent variable multivariate regression. Journal of Chemometrics, 10(1):31–45, 1996.

(317e) Quantitative Structure Retention Relationship (QSRR) Models in An Analytical Quality By Design Framework: Simultaneously Accounting for Compound Properties, Mobile Phase Conditions and Stationary Phase Properties

Muteki, K., Pfizer Inc
Morgado, J. E., Pfizer Inc
Reid, G. L., Pfizer Worldwide Research and Development
Wang, J., Pfizer Inc
Xue, G., Pfizer Inc
Riley, F., Pfizer Inc
Harwood, J. W., Pfizer Inc
Fortin, D. T., Pfizer Inc
Miller, I. J., Pfizer Inc
Originally presented on: 11/5/2013 14:10:00 - 14:35:00

Abstract

Quantitative Structure Retention Relationships (QSRR) can play an important role to enhance the speed and quality of chromatographic method development. This paper presents a novel (compound classification based) QSRR modeling strategy which simultaneously accounts for analyte properties, mobile phase conditions and stationary phase properties. It involves the adoption of two models: (A) Partial Least Square - Discriminate Analysis (PLS-DA) to classify compounds into sub-classes having similar interactive relationships between the mobile phase conditions and stationary phase; (B) L- Partial Least Squares (L-PLS) to predict the compound's retention time based on the mobile phase conditions, stationary phase, and the compound properties. For the retention time of a compound to be modeled, the most favorable compound class is identified in an optimization framework which simultaneously minimizes both the compounds misclassification rate (based on PLS-DA) and the retention time prediction error (based on L-PLS) through a Mixed Integer Optimization. The proposed QSRR model (L-PLS with compound classification) significantly improves the retention time predictability compared with traditional QSRR models or L-PLS models without compound classification. When combined with the Linear Solvation Energy Relationship (LSER) parameters (using Abraham coefficients) as the column properties, the approach allows: (1) prediction of (new, never analyzed) compound retention times under chromatographic conditions (columns and mobile phase conditions) used to train the model; (2) prediction of (previously analyzed under training conditions) compound retention times under chromatographic conditions that have not been previously evaluated; (3) optimization of the chromatographic conditions (mobile phase and column selection) to maximize critical pair resolution, including new compounds; (4) enhanced mechanistic understanding of the interactive retention relationship between compounds, mobile phase and column (e.g. compound retention mechanism). The effectiveness of proposed modeling strategy will be demonstrated through two practical pharmaceutical applications in Supercritical Fluid Chromatography (SFC) and Reversed Phase Liquid Chromatography (RPLC).

(317f) Case Study: Model-Based Chromatography Process Development

Hunt, S., Amgen Inc.
Todd, R.,
Belval, T.,
Tolley, K.,
Larsen, T.,
Originally presented on: 11/5/2013 14:35:00 - 15:00:00

The biopharmaceutical industry continues to explore options for improving process development efficiency to meet the needs of today's fast and competitive market. Technologies like platform utilization, high throughput screening, and computer simulation continue to advance the industry, but a strategy that successfully balances process efficiency with speed to market will likely require a synergistic application of these technologies. In recent years, the sophistication of the general rate model for chromatography has advanced considerably. Utilizing these models for chromatography process development lends itself favorably to high throughput isotherm measurement techniques. We present a case study for the application of a model-based approach to chromatography process development and characterization for a commercially relevant monoclonal antibody.

(343a) Improving the Robustness of a Pharmaceutical Crystallization Process

Marchut, A. J., Abbott Laboratories
Nordstrom, F. L., AbbVie
Patel, K.,
Bhagavatula, L., AbbVie
Originally presented on: 11/5/2013 15:15:00 - 15:40:00

Improving the Robustness of a Pharmaceutical Crystallization Process

Alexander Marchut, Fred Nordstrom, Ketan Patel, and Lakshmi Bhagavatula

Process Research and Development (Process Engineering, Solid State Chemistry, and Process Chemistry), Manufacturing Science and Technology

Abbvie

1401 Sheridan Road

North Chicago, IL 60064

Crystallization is often a key component of any pharmaceutical process – the desired active ingredient or intermediate is separated from the solution phase into a solid phase with the desire that impurities remain in the solution phase and the resulting solid particles have reproducible physical properties such as solid form, crystal habit, and particle size. Pharmaceutical crystallization processes must be robust so that the resulting solid phase properties are reproducible and independent of scale. We employed Quality by Design techniques to improve a pharmaceutical crystallization resulting in a more reproducible process with a broad design space. Our changes to the process were subject to certain regulatory constraints which will be discussed.

(343b) Development of An API Micronization Design Space Using a Design of Experiments Approach

Chekal, B. P., Pfizer Global Research & Development
Steeno, G. S., Pfizer Global Research & Development
Originally presented on: 11/5/2013 15:40:00 - 16:05:00

The design space for a micronization process used to reduce the particle size of a low solubility active pharmaceutical ingredient (API) has been developed using a resource-efficient D-optimal design of experiments (DoE) study. Micronization trials were completed using an Isopak-2 SuperJet Microniser (APTM, Switzerland). An experimental plan was designed to examine the impact of three micronization process parameters; feed rate, mill pressure, and Venturi pressure, on the D(v,0.5) and D(v,0.9) particle size averages. A broad range of the parameter space was examined in the DoE study, but with an additional imposed constraint limiting the difference between the mill pressure and the Venturi pressure to guarantee proper micronizer performance. Due to the irregular polygonal shaped parameter space caused by the imposed constraint, a 10-run D-optimal design was constructed to ensure both adequate coverage across the experimental region and precise modeling capability.

Statistical models generated from the measured D(v,0.5) and D(v,0.9) values demonstrated that these particle size averages have a strong dependence on the mill pressure and a slight dependence on the Venturi pressure. The feed rate had no measurable effect on the particle size averages, implying latitude with that parameter. Three subsequent micronization trials were run to independently verify the model predictions for the D(v,0.5) and D(v,0.9) particle size averages. Target ranges for the particle size averages were established with consideration of bioavailability, content uniformity, and drug product manufacturability. Due to its low aqueous solubility, the particle size of the API is limited to D(v,0.5) of no more than 8 microns and D(v,0.9) of no more than 17 microns to ensure bioavailability. By comparing the model predictions with these target ranges, the design space for the micronization process was identified with the constraint of no less than 2.0 bar for the mill pressure.

Use of a D-Optimal DoE for the efficient development of a robust API micronization design space with parameter constraints has been demonstrated as an alternative to traditional micronization development studies.

(343c) Model Based Design of Experiments for Pharmaceutical Process Development

Albrecht, J., Bristol-Myers Squibb Co.
Li, J., Bristol-Myers Squibb Company
Originally presented on: 11/5/2013 16:05:00 - 16:30:00

Design of experiments (DoE's) are widely used experimental methods to build process understanding. However, the classical design of experiments methodology is optimal only for systems with linear responses and where no prior knowledge exists for the system. Classical factorial designs are common within pharmaceutical drug substance process development despite the fact that these processes are non-linear in nature (i.e. temperature, chemical kinetics). Model based design of experiments allows the researcher to maximize, based on any prior knowledge, the information content of a future series of experiments. Before drug substance DoE's are planned, knowledge of the process is available from, at a minimum, route scouting experiments. Even for reactions where the mechanism is not known, credible options can often be proposed from first principles. From the proposed mechanism(s), a family of models can be drafted. Thus, the combination of any amount of data with the experience and judgment of the researcher can be codified into a model more appropriate than the "no prior knowledge" DoE assumption.

A number of model-based experimental design methodologies have been developed that address the needs of experimental design within the pharmaceutical industry. This paper will report model based design of experiments applied to drug substance synthesis and include examples of optimal experimental design ranging from identifying under-explored experimental conditions to Bayesian updating to refine reliable process operating regions.

(343d) Application of in-Situ Raman Spectroscopy in Monitoring the Dynamic Trends of Polymorphic Transformation and Hydrogenation Reaction

Zhang, H., Sunovion Pharmaceuticals Inc.
Prytko, R. J., Sunovion Pharmaceuticals Inc.
Sizensky, M.,
Saranteas, K., Sunovion Pharmaceuticals
Originally presented on: 11/5/2013 16:30:00 - 16:55:00

Application of in-situ Raman spectroscopy in monitoring the dynamic trends of polymorphic transformation and hydrogenation reaction

Haitao Zhang, Bob Prytko, Mike Sizensky, Kostas Saranteas

Chemical Process Research & Development, Sunovion Pharmaceuticals Inc., 84 Waterford Drive, Marlborough, Massachusetts 01752, U.S.A.

Abstract

During the last decade many novel manufacturing practices such as Process Analytical Technology (PAT), Quality by Design (QbD), and Real Time Release (RTR) have been introduced to the pharmaceutical industry, and they are expected to be an effective aid to pharmaceutical manufacturing. On the basis of a recent FDA directive, this mode of operation will transform the manufacturing aspect of the pharmaceutical industry.

In crystallization processes, control and optimization of the polymorph formation requires in-situ quantitative measurement of the concentration of different polymorphs. Herein, we report the application of in-line Raman spectroscopy to monitor the polymorphic transformation of DSP-6373 during crystallization process. Kinetically preferred Form II of DSP-6373 was transformed into thermodynamically stable Form I during a solvent-mediated phase transition in isopropanol. The feasibility of identifying the differences between the two polymorphs has been demonstrated by collecting and comparing the Raman spectra of the pure polymorphic crystals. A calibration curve has been built based on different ratios of the polymorphs. The in-line Raman spectroscopy can effectively identify the two forms of crystals in solution as well, which allows it to be used to capture the dynamic trend of the polymorphic transformation.

This study experimentally demonstrates the advantages of using in-line Raman spectroscopy for monitoring in-situ pharmaceutical crystallization by detecting the polymorphic transformation, which is fundamental in the development and operation of industrial crystallization processes. A second application of in-line Raman spectroscopy technology was also explored for the hydrogenation reaction of an enacetamide to an acetamide in order to monitor the real time performance of the reaction and obtain in-line kinetic data.

(343e) Application of Combined Experimental and Modeling Approach to Understand pH Driven Reactive Crystallization

Vaidyaraman, S., Eli Lilly and Company
Conder, E., Eli Lilly & Co
Brown, G., Eli Lilly and Company
Stack, A.,
Phillips, M., Eli Lilly & Co
Viswanath, S., Eli Lilly and Company
Moher, E., Eli Lilly & Co.
Originally presented on: 11/5/2013 16:55:00 - 17:20:00

Application of Combined Experimental and Modeling Approach to Understand pH driven Reactive Crystallization

Authors: Shankar Vaidyaraman, Edward Conder, Gregory Brown, Angela Stack, Michael Phillips, Shekhar Viswanath, Eric Moher

Small Molecule Design and Development, Eli Lilly and Company, Indianapolis, IN 46285

This talk will discuss a combined experimental and modeling approach to understand pH driven reactive crystallization. A mechanistic approach was taken by using experiments to determine the fundamental parameters of the system (pKa, solubility product, pseudo-growth constant). A "fit for use" mechanistic model accounting for the different reaction equilibria in solution was used to link the fundamental parameters to process parameters (addition type, pH during crystallization, time duration of crystallization etc.). The model was used to help plan experiments to understand the impact of flow upsets on pH excursion during crystallization. The model was also used to generate a response surface of pH vs other process parameters to provide guidance to pH controller design.

(343f) Model Based Development and Validation of a Preparative Size Exclusion Chromatography Step

Hansen, E. B., Novo Nordisk
Hansen, T. B., Novo Nordisk A/S
Sejergaard, L., Novo Nordisk
Originally presented on: 11/5/2013 17:20:00 - 17:45:00

Model based development and validation of a preparative Size Exclusion Chromatography step

Ernst B. Hansen, Thomas B. Hansen and Lars Sejergaard Novo Nordisk A/S, Hagedornsvej 1, DK-2820 Gentofte, Denmark

The application of mechanistic models for process development and optimization has been a topic of discussion in the Pharmaceutical industry for many years. Mechanistic models are considered as an alternative to a standard DoE type empiric approach. Most often the DoE approach is preferred based on a cost-benefit evaluation where the time consumption and the complexity are found to extensive when using mechanistic modelling. In our company we use mechanistic models for the development of selected downstream process steps. A Model based approach is preferred when a suitable model has been established, in the literature or in-house, and the process step is considered a key step regarding economy and/or product quality.

The presented case deals with Size Exclusion Chromatography (SEC) as the final purification step for a therapeutic protein. The model was developed with the purpose of using simulation for the process evaluation which is done in preparation to the production scale *Process Performance Qualification* (PPQ). The main function of the SEC step is to reduce the content of larger proteins (HMWP and dimer). The model is based on differential mass balances for each component and includes axial dispersion and mass transfer as dispersion effects. The model was calibrated using four lab runs with variation in load, flow and bed height. The fitted model parameters for each component were the exclusion coefficient ($0<K_d<1$) which determines the retention in the column and the mass transfer coefficient related to dispersion (peak width). Before doing the actual process evaluation we had to deal with a challenge concerning the concentration in the product pool from the SEC step. The feed concentration to the SEC step varied quite a bit and since the SEC step can only dilute the product (no adsorption takes place) the variation in the feed concentration results in a variation in the pool concentration. As this is the final step before formulation there is a lower limit to the concentration which must be met. Using simulation we were able to create enough data to visualise the problem and come up with a new strategy for loading the column which can compensate for the variation in the feed concentration. The *process evaluation* is a systematic check of the performance of the step when varying the important process parameters within their operating range. Due to the experimental constraints (time, availability of product) this is often done as a reduced design (DoE). Since we were using simulation there was no need to set up a reduced design. The evaluation was thus based on 216 simulations covering the operating range of the important process parameters (e.g. load, flow, bed height, OD cut). A perhaps overlooked advantage of using simulation is that both the concentration and the content (impurity levels) of the feed solution can be varied as part of the study without any extra effort such as dilution, concentration, or spiking with impurities.

Our experience of using modelling is very positive. The initial extra energy put into understanding and developing the models and the methods for calibration is worthwhile. But, contrary to the general discussion of which approach is better, it is not a reduction in resources or time consumption that drives us. It is the level of process understanding that we gain from using a model based approach compared to an empiric approach. Knowing what is going on we raise the bar for what is expected and what is unexpected. When the model describes the data well we are confident that the process is running as would be expected. This can include non-linear and counter intuitive behaviour. On the other hand when there is a systematic descrepancy between the model and the data we acknowledge that something unusual is going on and start looking for the unexpected phenomena.

(344a) Manufacturing of Low Fill Weight Capsules Using a Nozzle Dosator Capsule Filling Machine

Llusá, M., Research Center Pharmaceutical Engineering GmbH

Faulhammer, E., Research Center Pharmaceutical Engineering GmbH

Fink, M., Research Center Pharmaceutical Engineering GmbH

Scharrer, G., RCPE GmbH

Calzolari, V., MG2

Biserni, S., MG2

Lawrence, S., GSK

Bresciani, M., Research Center Pharmaceutical Engineering GmbH

Khinast, J. G., Graz University of Technology

Originally presented on: 11/5/2013 15:15:00 - 15:40:00

Introduction

Capsule filling using nozzle dosators (ND) has been widely investigated (Jones, 2001; Newton, 2012) and it is one of the main technologies used by pharmaceutical industry today. Nozzles come in sizes that match the diameter of capsules (from 00 to 5), and deliver a specific volume of powder into capsules. However, the amount of powder that the smallest nozzle (size 5) can handle is in the range of 40 mg-80 mg (depending on the density of the powder formulation).

Some new and prospective pharmaceutical products require the handling of even smaller powder quantities: 5mg-20 mg. A straightforward solution is to extend the use of nozzle dosators (by reducing the diameter of nozzles) and take advantage of the features already available for this technology: large output, highly automated, check weight for quality control of every single capsule. However, this requires understanding the design space for nozzles of smaller diameter and to revisit the effects of process parameters and powder properties on capsule quality. It is expected that miniaturizing the nozzles will affect the process performance and capsule quality.

In the current research, Design of Experiments (DOE) as part Quality by Design approach is used to study the effects of process parameters and material properties for three small nozzles on fill weight and weight variability. A capsule filling machine (Labby, MG2) with a single ND is used.

Materials and Methods

The DOE (D-Optimal with Design statistics G-Efficiency) includes four process parameters of the capsule filling machine (Labby, MG2): 1- Diameter of the nozzle (3 different nozzles), 2- Size of dosing chamber (three sizes), 3- Depth of powder bed in the bowl (3 values), and 4- Speed of capsule filling (3 values). The DOE also considers two interactions between process parameters: 1- Size of dosing chamber and powder layer depth; and 2- Diameter of the nozzle and size of dosing chamber. The DOE has two constraints for the ratio between size of dosing chamber and layer (interaction 1): never smaller than 1:2 and never larger than

1:5 (which are typical values found in capsule filling machine set-up).

The materials are four different grades of lactose: Respitose ML001, ML006, SV010, and SV003 (DFE). The following material attributes are characterized (in triplicates): particle size and shape; bulk and tapped density; compressibility; 3 flow indexes (Carr index, Basic flowability energy (measured with FT4), and ffc (measured with FT4)); cohesion (measured with FT4), wall friction, and air permeability (measured with FT4).

In order to precisely assess capsule fill weight, and hence guarantee the validity of the conclusions of the current study, we follow the following procedure:

1. Weighing of empty capsules: The powder is filled into size 3 gelatin capsules, which are approximately 48 mg ± 3 mg. The empty body of capsules is much heavier than the powder fill itself. The variability in the weight of the empty capsules (± 3 mg) is almost as large as the low-dose content (5 mg). Therefore, the weight of the empty body must be known before filling capsules. Empty capsules are weighed and numbered for identification before filling.
2. Use a high precision scale in our experiments: DENVER SI-234A (Readability: 0.0001 g; Reproducibility: 0.1 mg).

For each run in the DOE, capsule filling is performed with MG2 equipment (Labby) and the method is as follows:

1. Fix the size of the dosing chamber with the aid of adjustment gauges (2.5mm, 3.75mm, and 5mm).
2. Fill capsules for 5 minutes to reach steady state of the operation.
3. Collection and weighing of the first group of 20 numbered capsules.
4. Continuously run the machine for 5 minutes.
5. Collection and weighing of the second group of 20 numbered capsules.
6. Compare the net weight of the two groups of capsules.
7. If the fill weight of the two groups of capsule does not differ by more than 10%, then the data is combined in a single group (a batch of 40 capsules).

Results

First, the effects of four process parameters and two of their interactions on fill weight and weight variability for four powders is discussed. The lines in Table 1 corresponds to each run in the DOE and show the values for each of the four process parameters, and the fill weight and weight variability for each of the four powders tested. At the bottom of the columns named weight and weight variability for each powder, we indicate the parameters that significantly correlate with each of these two capsule attributes. For instance, fill weight always correlate with the diameter (dia) of the nozzle and the size of the dosing chamber (cha). Hence, filling is performed on volumetric basis.

At the bottom of the columns for weight variability, the parameters that affect fill weight variability are indicated. These parameters depend on the type of lactose tested and nozzle diameter (dia) is always one of them. The interactions between process parameters do not significantly affect either fill weight or weight variability.

					RESPITOSE ML001		RESPITOSE ML006		RESPITOSE SV010		Respitose SV003	
RUN	speed	diameter dosater	chamber	powder layer	weight	weight variability	weight	weight variability	weight	weight variability	weight	weight variability
1	2500	1,9	2,5	5	5,25	6,73	4,19	10,70	6,11	4,80	5,96	7,13
2	1500	2,8	3,75	10	23,77	2,71	14,77	10,43	19,97	1,87	19,11	1,78
3	2500	3,4	2,5	12,5	45,49	1,07	23,04	2,06	22,44	2,48	21,45	3,56
4	500	3,4	2,5	5	21,27	7,75	18,26	4,54	19,78	1,92	19,49	2,67
5	500	3,4	2,5	12,5	26,64	2,30	24,72	1,59	21,20	2,63	20,59	1,87
6	500	1,9	5	12,5	11,39	5,48	10,14	5,82	11,04	3,20	11,53	3,09
7	2500	1,9	2,5	12,5	7,28	5,30	6,93	5,02	6,82	3,93	6,66	6,81
8	2500	3,4	5	12,5	45,65	1,95	35,85	4,42	38,72	1,26	37,11	1,13
9	2500	1,9	5	10	10,68	11,44	7,35	9,20	11,21	4,25	10,63	4,75
10	1500	2,8	3,75	10	23,06	2,11	14,87	9,37	19,87	1,77	19,22	2,82
11	1500	2,8	3,75	10	22,32	3,66	11,85	8,79	19,73	2,54	19,12	2,62
12	2500	3,4	2,5	5	19,60	7,21	14,10	5,00	20,96	2,02	19,81	3,43
13	500	1,9	2,5	5	6,43	4,84	4,77	5,72	6,10	4,54	6,33	6,80
14	500	3,4	5	10	43,38	3,37	33,59	3,77	37,36	1,55	36,27	1,43
				FACTORS	dia/cha	dia/lay	dia/cha/lay	dia	dia/cha	dia/cha	dia/cha	dia

Table 1: Effect of process parameters on fill weight and weight variability for four different grades of lactose (Respitose ML001, ML006, SV010, and SV003).

Then, the effects of material attributes (for the group of four powders) on fill weight and weight variability is discussed. Figures 1 and 2 show the material attributes that have a statistically significant correlation with fill weight and weight variability respectively. Bulk and tapped density have the largest correlation with fill weight. Many material attributes affect weight variability, however their effect is rather small (see small coefficients on y axis of Figure 2). Particle size shows the largest correlation with weight variability.

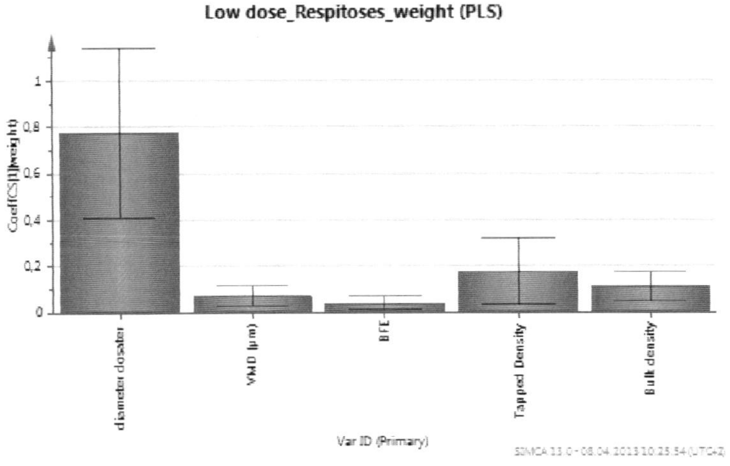

Figure 1: Material attributes that affect fill weight. Bulk and tapped density show the strongest correlation with fill weight.

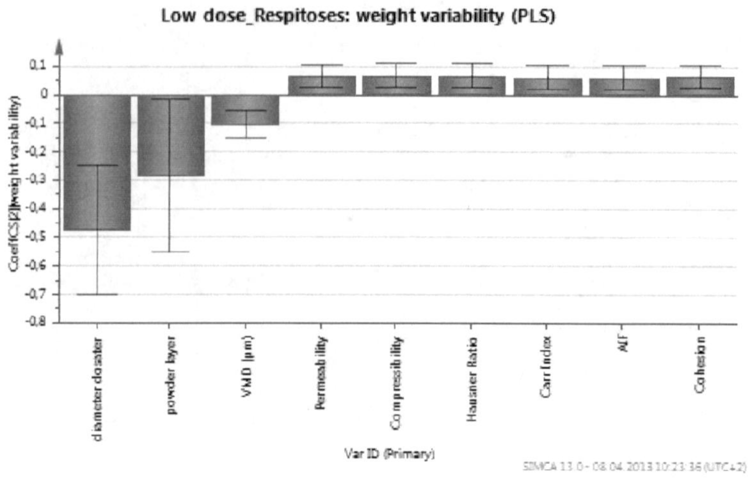

Figure 2: Material attributes that affect fill weight variability. Many parameters have a small effect but still statistically significant. Particle size has the largest effect on weight variability.

Conclusions

The current study uses QbD principles to understand the effect of process parameters of a nozzle dosator capsule filling machine and material attributes on fill weight and weight variability of capsules with low fill content. This study is focused on low fill weight (5mg-45mg). In order to fill these amounts of powder, nozzles with smaller diameters than the traditional nozzles (sizes 0 to 5) are required.

For the powders tested, fill weight is affected by the diameter of the nozzle and the size of the dosing chamber. Diameter and size of the chamber are the two parameters that define the volume of the dosing chamber (the R^2 of models is at least 90%). Hence, filling is performed on volumetric basis.

The material attributes that have the largest effect on fill weight are bulk and tapped density. Many material attributes affect weight variability, even though their effect is rather small.

Acknowledgments

To MG2 and GSK for their financial and scientific support

References

Jones, B.E. (2001). The filling of powders into two-piece hard capsules. International Journal of Pharmaceutics 227, 5-26.

Newton, J.M. (2012). Filling hard gelatin capsules by the dosator nozzle system - Is it possible to predict where the powder goes?. International Journal of Pharmaceutics 425, 73-74.

(344b) Drug-On-Demand: A Mini-Manufacturing Method Using Drop-On-Demand Technology

Hirshfield, L., Purdue University
Reklaitis, G. V., Purdue University
Giridhar, A. V., Purdue University
Harris, M. T., Purdue University
Originally presented on: 11/5/2013 15:40:00 - 16:05:00

In recent years, the FDA has encouraged pharmaceutical companies to place renewed emphasis on model-based process design and the development of more innovative, efficient manufacturing methods. "Mini-manufacturing" of medicine can alleviate the need to mass-manufacture all drugs and can be beneficial in the creation of high-potency drug forms or products tailored to a specific patient. We describe one such mini-manufacturing process, "drug-on-demand," which uses drop-on-demand printhead technology to deposit API onto edible substrates. We use a high precision positive displacement pump to deposit solvent-based drug systems or polymer-drug melts. This method allows for the creation of individualized dosage forms, multi-layered films with all of the API needed for one patient, or the creation of personalized medicine, in which the amount of API can be varied depending on the patient.

In this work, we present the details of the drug-on-demand pilot facility created as part of the Engineering Research Center for Structured Organic Particulate Systems. We present a process narrative of the setup and discuss the automation implemented on the drug-on-demand system. The system, consisting of a positive-displacement pump, xy staging, imaging system, and various temperature controllers, allows for precise control over the material properties, drug morphology, drop size, and drop dynamics. With careful consideration of process parameters and an effective control strategy, we can execute an automated, optimized, and controlled print cycle while closely monitoring drop size, drug morphology, and drop deposition pattern. In this paper, we summarize our achievements and current research in this area, including a description of the engineering principles behind our process and analysis of drug forms created using solvent-polymer systems.

(344c) Supervisory Control of a Drop-On-Demand Mini-Manufacturing System for the Production of Pharmaceuticals

Icten, E., Purdue University
Nagy, Z. K., Purdue University
Reklaitis, G. V., Purdue University
Originally presented on: 11/5/2013 16:05:00 - 16:30:00

The support of the FDA in the development of more efficient production technologies creates new possibilities for innovation in pharmaceutical production processes. There is an intention within the pharmaceutical industry to move from traditional batch processing to continuous manufacturing. [1] As a part of the Engineering Research Center for Structured Organic Particulate Systems, a mini-manufacturing process for drug formulation has been built. The process utilizes the drop-on-demand (DoD) printing technology for predictable and highly controllable deposition of active pharmaceutical ingredients (API) onto an edible substrate, using a semi-continuous operation suitable for low volume production of personalized dosage forms.[2] The system consists of a positive displacement pump, xy-staging, camera and temperature controllers.

Here, we present a supervisory control framework for individual dosage forms with precise control of drop size, morphology and formulation composition. Implementing a supervisory control system on the mini-manufacturing process, including automation and closed-loop control, is essential for producing individual dosage forms with the desired properties. The drop volume is monitored using the imaging system, to ensure consistent drop size. The xy-staging allows creating precise drop positioning while printing.

Using this process, different drug formulations including solvent-polymer-API and polymer-API systems, i.e. co-melt systems, can be produced. Polymers are added to the printing material to help control drug morphology, material properties and the formulation composition.[3] In order to produce melt-based dosage forms the complete system must be heated. Therefore we implemented temperature control around the reservoir, around the tubing and within the pump. Since crystallization temperature will have an effect on product quality, influencing the bioavailability of the drug, precise temperature control on the nozzle and substrate is important.[4] After drop deposition, changes in drug morphology can be monitored using Near Infrared spectroscopy and corrected by temperature control action on the substrate. Using the proposed temperature control strategy on the substrate, one can tailor the crystallization behavior to compensate for these variations and achieve consistent drug morphology.

1. K. V. Gernaey, A. E. Cervera-Padrell, and J. M. Woodley, "A perspective on PSE in pharmaceutical process development and innovation," *Computers & Chemical Engineering*, vol. 42, pp. 15–29, Jul. 2012.
2. B. Derby, "Inkjet Printing of Functional and Structural Materials: Fluid Property Requirements, Feature Stability, and Resolution," *Annual Review of Materials Research*, vol. 40, no. 1, pp. 395–414, Jun. 2010.
3. N. S. Trasi and L. S. Taylor, "Effect of Additives on Crystal Growth and Nucleation of Amorphous Flutamide," *Crystal Growth & Design*, vol. 12, no. 6, pp. 3221–3230, Jun. 2012.
4. M. Fujiwara, Z. K. Nagy, J. W. Chew, and R. D. Braatz, "First-principles and direct design approaches for the control of pharmaceutical crystallization," *Journal of Process Control*, vol. 15, no. 5, pp. 493–504, Aug. 2005.

(344d) Application of Johanson Model Calibrated With Instrumented Roll Data for Roller Compaction Process Development and Scale Up

Nesarikar, V., Bristol Myers Squibb

Originally presented on: 11/5/2013 16:30:00 - 16:55:00

Roller compaction is a dry granulation process used to convert powder blends into free flowing agglomerates. During scale up or transfer of roller compaction process, it is critical to maintain comparable ribbon densities at each scale in order to achieve similar tensile strengths and subsequently similar particle size distribution of milled material. Similar ribbon densities can be reached by maintaining analogous normal stress applied by the rolls on ribbon for a given gap between rolls. Johanson (1965) developed a model to predict normal stress based on material properties and roll diameter. However, the practical application of Johanson model to estimate normal stress on the ribbon is limited due to its requirement of accurate estimate of nip pressure i.e., pressure at the nip angle. Another weakness of Johanson model is the assumption of a fixed angle of wall friction that leads to use of a fixed nip angle in the model.

To overcome the above mentioned limitations, a novel approach was developed using roll force equations based on a modified Johanson model in which the requirement of pressure value at nip angle was eliminated. An instrumented roll on WP120 roller compactor (See Figure 1) was used to collect normal stress data (P1, P2, P3) measured at three locations across the width of a roll, as well as gap and nip angle data on ribbon for placebo and various active blends along with corresponding process parameters. The sensor # 1 is located towards the inner edge of the roll and records normal stress P1, the sensor # 3 is located towards the outer edge of the roll and records normal stress P3, and the middle sendor records normal stress P2.

The nip angles were estimated directly using experimental pressure profile data of each run. Example of nip angle calculation for one of the placebo runs is provided in Figure 2. The angle between intersection of tangents to ascending and descending portions of pressure peak with the base line was used as a nip angle.

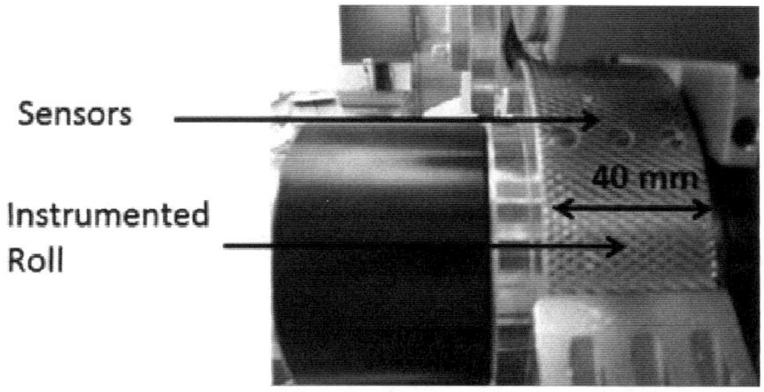

Figure 1: Instrumented Roll Installed on Alexanderwerks® WP120 roller compactor

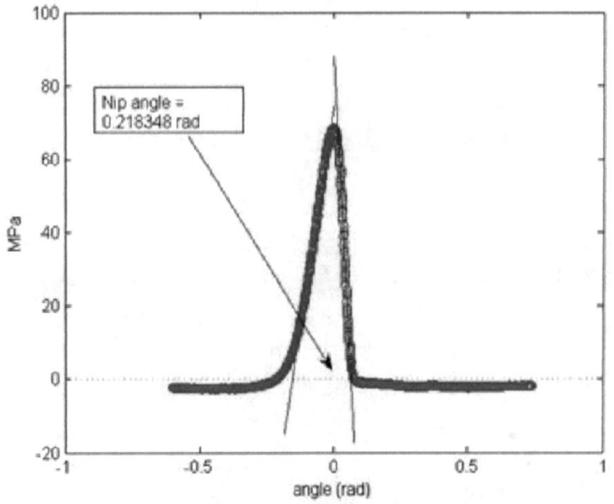

Figure 2: **Nip Angle Calculation for Placebo Pressure Profile (Roll Speed = 4 rpm, Screw Speed = 19 rpm and Roll Pressure = 40 bar, and Gap = 2.65 mm)**

The equations (1) and (2) below show the roll force (RF) as a function of gap (S), nip angle (α), roll diameter (D), roll width (W), and pre blend compressibility (K).

$$RF = \frac{Pm * W * D * F}{2} \quad ..... \quad Eqn\ (1)$$

Pm (i.e. P2) is maximum normal stress at the center of the roll width.

And F is defined as

$$F = \int_0^{\alpha} \left[\frac{\left(\frac{S}{D}\right)}{\left(1 + \frac{S}{D} - cos\theta\right) cos\theta} \right]^K cos\theta d\theta \; ......Eqn\ (2)$$

$$RF(kN) = \frac{0.86 \times HP(bar)}{(10\ bar)(cm\ roll\ width)} \times (W) \quad ...\ Eqn\ (3)$$

Roll diameter (D) is 120 mm for WP120 compactor. The compressibility (K) is determined from the reciprocal of the slope of initial linear portion of the logarithmic plot of density as a function of pressure data obtained in uniaxial compaction. The roll force equation of Johanson model was validated using normal stress, gap, and nip angle data of the placebo runs.

The calculated roll force values (i.e. Johanson model) compared well with those determined from the vendor supplied roll force equation (3) provided for the Alexanderwerk® WP120 roller compactor.

Subsequently, the calculation was reversed to estimate normal stress (Pm or P2) and corresponding ribbon densities as a function of gap and roll pressure. A set of calibration runs are conducted for a given preblend using 2^2 factorial with gap and roll pressure as two factors. In absence of such data, a set of runs with different combinations of gap and roll pressure can be used. A calibration set of 3 or more runs is ideal for this purpose.

The purpose of calibration runs are as follows. In our research (Nesarikar et al. 2012), we have shown that the normal stress values recorded by side sensors P1 and P3 were lower than normal stress values recorded by the middle sensor P2. This is attributed to heterogeneity of feeding pressure in the last flight of the feed screw and this has also been previously reported in the literature (Simon et al. 2003). In addition, using FEM simulations, Cunningham et al. (2010) also showed that the normal stress and relative density decreased near the ribbon edges due to side seal friction, resulting in variation of relative density across ribbon width.

The variation of normal stress across ribbon width leads to variation of ribbon density. Use of Johanson model without taking into account the normal stress variation can over predict the ribbon density as ribbon density will be calculated using Pm (or P2) alone. Therefore, modeling approach we used takes into account the normal stress variation across ribbon width to avoid over prediction of ribbon densities.

The typical assumed normal stress profile across ribbon width for a roller compactor with a single feed screw system is shown in Figure 3. As shown in Figure 3, Pe is normal stress at the ribbon edge. x1 is the distance from the center on both sides over which P2 (i.e Pm) is effective. The experimental ribbon density data of calibration set is used to determine x1 and Pe values. Note that sensors 1 and 3 are located 6.75 mm each from the roll edge. True density and pressure-porosity data of pre blend is used to calculate densities corresponding to normal stress at each location on the ribbon. Once density profile across ribbon width is obtained, trapezoidal rule is used to estimate area under curve for calculating average ribbon density.

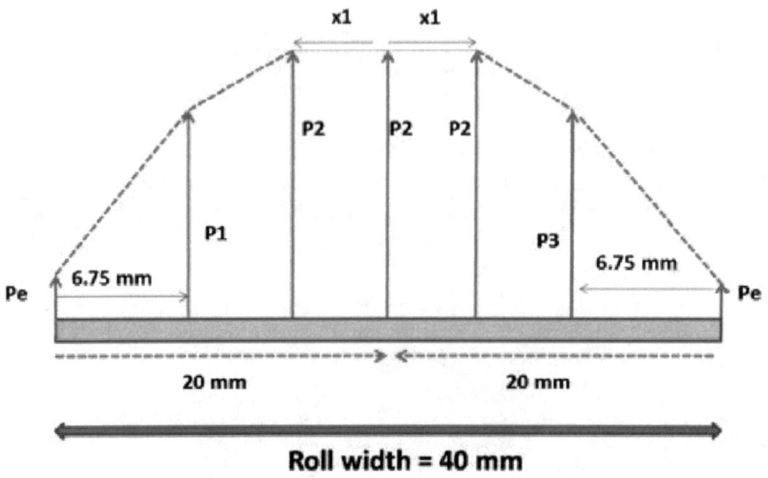

Figure 3: Assumed pressure profile across roll width on WP120 with distance (x1) and normal stress (Pe) at the edge optimized to match experimental ribbon density.

The model predicted ribbon densities of the placebo runs compared well with the experimental data. The placebo model also predicted with reasonable accuracy the ribbon densities of active A, B, and C blends prepared at various combinations of process parameters. Example of prediction for ribbon densities of active B batches is shown in Figure 4. Active B was a low drug load (<= 5% w/w) formulation.

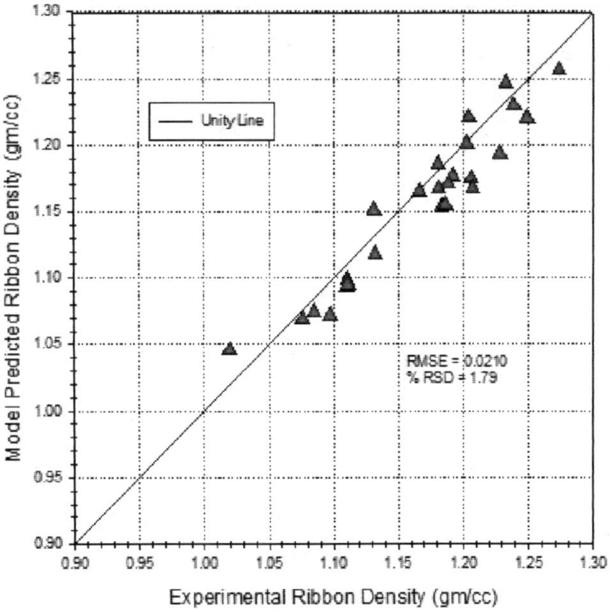

Figure 4: Prediction of Ribbon Densities of Active [B] Batches using Placebo Model

We have also successfully demonstrated the extension of this model to scale up from Alexanderwerk® WP120 to WP200 unit. The scale up approach presented in this work is not limited to WP200 and can be applied to any other roller compactor since the model inputs; gap and RFU are machine independent parameters. Example of scale up from Alexanderwerks WP120 to WP200 for the active C batches is shown in Table 1.

Table 1: Model Predicted vs. Experimental Ribbon Densities of Active [C] Scale up Batch 2 Prepared on WP200

Run	Roll Pressure (bar)	Gap (mm)	Model Predicted Ribbon Density (gm/cc)	Experimental Ribbon Density (gm/cc)
1	60	2.2	1.0748	1.0526
2	60	2.8	1.0538	1.0336
3	75	2.5	1.1027	1.0819
4	90	2.2	1.1415	1.1198
5	90	2.8	1.1236	1.1110

RMSE = 0.019815

References

1) Cunningham J., Winstead D., Zavaliangos A. 2010. Understanding variation in roller compaction through finite element-based process modeling. Computers and Chemical Engineering. 34, 1058-1071.

2) Johanson, J. R., 1965. A rolling theory of granular solids. ASME, J. Applied Mechanics Series E 32 (4), 842-848.

3) Nesarikar V, Vatsaraj N., Patel C., Early W., Pandey P., Sprockel O., Gao Z., Jerzewski R., Miller R., Levin M. Instrumented Roll Technology for the Design Space Development of Roller Compaction Process. International Journal of Pharmaceutics. 426 (2012) 116-131

4) Nesarikar V, Patel C., Vatsaraj N., Early W., Pandey P., Sprockel O., Jerzewski R., Roller compaction process development and scale up using Johanson model calibrated with instrumented roll data. International Journal of Pharmaceutics. 436(2012) 486-507.

5) Simon O. and Guigon P., 2003, Correlation between powder-packing properties and roll press compact heterogeneity, Powder Tech. 130, 257-264.

(344e) Parametric Analysis of Fluidized Bed Drying of Pharmaceutical Materials

Liu, X., Rutgers University
Muzzio, F., Rutgers University
Khinast, J. G., Graz University of Technology
Glasser, B., Rutgers University
Originally presented on: 11/5/2013 16:55:00 - 17:20:00

Fluidized beds are extensively used in the pharmaceutical and other chemical industries either as a batch or continuous process for drying moist powders and granular solids because of good mixing of solids and intensive heat and mass transfer between the solid and hot gas phases in the system. Generally, high energy input is required during this process to provide the latent heat of water evaporation, and thus on-line measurements become important to determine the optimal operation conditions in order to minimize energy usage. In the pharmaceutical industry, the drying step in many pharmaceutical production lines is often a bottleneck because it is difficult to determine the time required to dry the materials as well as analytically verify that a predetermined end point has been reached. If the active pharmaceutical ingredient is heat sensitive or has a low melting point, controlling the drying process becomes more critical due to the sensitivity of the formulation to the temperature and the risk of compromising product quality and performance during drying.

In this work, two dibasic calcium phosphate anhydrous (DCPA1 and DCPA2) species with different particle sizes and porosities, have been examined using a Glatt GPCG-5 spray fluidized bed dryer/granulator and a Mini- Glatt spray fluidized bed dryer/granulator. The variations of the pressure drop across the bed, the product temperature and the humidity in the dryer air are measured as a function of time. The moisture content of the product during drying is measured using two approaches: 1) LOD - Taking samples from the bed every 5 to 10 minutes, drying the samples in an oven and calculating the sample moisture based on the mass loss before and after drying. 2) Using NIR to on-line measure the sample moisture. We have compared the NIR predictions and the LOD measurements, and a good match can be obtained. This indicates that NIR is a reliable approach to determine the end point of the drying process.

Parametric analysis is carried out based on two-factorial tests. We have examined the impact of drying air temperature, drying air flow rate, material loading, and initial moisture content on the end point of the drying time. From the statistical analysis we found that the four variables used in the factorial tests are independent and the most significant variables are drying air flow rate and initial moisture content. If the initial water volume is above the total pore volume of the particulate materials, drying can be divided into three stages. At the beginning of drying, water in the bed voids is evaporated (Stage I). Once all the water in the bed voids is removed, convection drives the water inside the particle pores to flow towards the particle surface, and then water evaporation occurs at the surface (Stage II). With further drying, the effect of convection reduces due to the loss of water and the drying front starts to penetrate inside the pores (Stage III). In stage I and II, drying is dominated by a constant drying rate stage with a constant bed temperature. In stage III, the drying process transits from a constant rate stage to a falling rate stage, and the bed temperature goes up. Drying is mainly dominated by the constant drying rate stage with zero order kinetics if drying is carried out at a relatively slow drying regime which is required by many pharmaceutical productions. A comparison of these two Glatt dryers and an evaluation of energy efficiency are also carried out in this work. Our work serves to characterize performance of fluidized bed dryer systems and provide physical insight into the fundamentals of drying of porous pharmaceutical materials.

(344f) Linking Powder Bed Heterogeneities to Tableting Performance

Akseli, I., Boehringer Ingelheim
Ladyzhynsky, N., Boehringer-Ingelheim
He, X., Boehringer Ingelheim
Deanne, R., Boehringer Ingelheim
Originally presented on: 11/5/2013 17:20:00 - 17:45:00

Tablet compaction presses and compaction emulator were used to rank order capping tendencies of a diverse sample set. The compression forces of 5 kN, 10 kN, 15 kN, 20 kN, and 25 kN were studied for oval and round tooling shapes, while 4 kN, 6 kN, 8 kN, 10 kN, and 14 kN were chosen for the beveled edge due to the force restrictions for this tooling. Compaction speeds of 25 rpm (24.8 ms dwell time), 40 rpm (15.5 ms dwell time), and 80 rpm (7.9 ms dwell time) were chosen as representative ranges for bench-to-pilotscale processing. Tablets were tested by in-house developed nondestructive ultrasonic method to obtain effective elastic (E) and shear (G) moduli in the axial and radial directions. The anisotropic properties of tablet formulations are represented by a dimensionless ratio (EG). The ultrasonic measurements revealed that effective elastic and shear modulus values vary with different testing orientations which indicated elastic and shear moduli anisotropy in tablets. It was found that as the EG ratio decreases; the compressed powder exhibited an increasing level of anisotropic nature, which led to a higher tendency for capping. Acceptance criteria of the EG ratio have been proposed based on manufacturing performance. Predictive computational compaction modeling was used. It was shown that altering process conditions such as tooling shape, compression force and press speed significantly affect the capping tendency of tablet formulations. A systematic approach has been applied to develop predictive tools to assess capping tendencies of formulations. These tools have been shown to correlate well with actual capping performance of a diverse sample set. The developed tools are fast, material sparing and have potential to flag the risk of capping issues and provide insight into the performance of common materials during formulation and process development. Application of these tools may help reduce development time, API usage and facilitate development of a robust product.

(404a) Application of QbD Principles for Tablet Coating Process Development, Scale-Up, and Transfer

Chen, B., AbbVie
Anderson, B. C., AbbVie
Originally presented on: 11/6/2013 8:30:00 - 8:55:00

Film coating is extensively used in the pharmaceutical industry for either functional or aesthetic purposes. In addition to numerous empirical approaches, some mathematical models have been reported to predict coating performance in various coaters. However, most of these models are limited to the thermodynamic aspect of the process. The kinetics/dynamics of the coating process is believed to play an equally important role in determining the final product quality. Kinetic/dynamic models can also ensure that the content uniformity of the coating film is achieved.

The combination of a dynamic/kinetic model and a thermodynamic model ensures consistent product quality being maintained across different equipment and/or different manufacturing scales. Such mythology has been shown to be an effective tool to enable successful tablet coating process development, scale-up, and transfer with minimal experimentation. Case studies will be presented to highlight the broad applications of this scale-independent and equipment-independent approach for the film coating process development.

(404b) Overall Elegance Assessment for a Solid Oral Dosage Form Using Multivariate Image Analysis and Multivariate Texture Analysis

Ottavian, M.,
Garcia-Munoz, S., Pfizer Worldwide R&D
Barolo, M., University of Padova
Originally presented on: 11/6/2013 8:55:00 - 9:20:00

The use of Multivariate Image Analysis (MIA) and Texture Analysis (MTA) has been proposed to analyze the elegance of materials involved in the manufacture of solid oral dosages, including the final product (García-Muñoz, Gierer 2010;Garcia-Munoz, Carmody 2010). This contribution discusses the application of these tools to support the design of the film coating step for a controlled release matrix tablet to assure the within-tablet color uniformity and prevent the logo erosion in the core. The embossing in the tablet required the application of additional algorithmic solutions, the *fast template matching* (Lewis 1995) (originally applied in the production of the movie "Forrest Gump") was used to separate the embossed logo from the rest of the tablet. Metrics were developed and correlated back to processing conditions in the film coating process. The obtained model was then used to suggest operational conditions to assure the elegance of the tablet and the lack of erosion in the face of the tablet. Additional details on the maintenance of MIA and MTA systems (not discussed in literature) are addressed, specifically dealing with changes in the lighting conditions. The application of the dynamic time warping algorithm to the 2D score histogram is proposed as a solution to the non-linear bias introduced by these changes.

References

García-Muñoz, S., Gierer, D. S. (2010). Coating uniformity assessment for colored immediate release tablets using multivariate image analysis.*International Journal of Pharmaceutics*, *395*(1-2),104.

Garcia-Munoz, S., Carmody, A. (2010). Multivariate wavelet texture analysis for pharmaceutical solid product characterization.*International Journal of Pharmaceutics*, *398*(1-2),97.

Lewis, J. (1995). Fast template matching.*Vision Interface 95, Canadian Image Processing and Pattern Recognition Society, Quebec City, Canada, May 15-19*,120.

(404c) Rapid Process Development of Chromatographic Process Using Direct Analysis in Real Time Mass Spectrometry As a Process Analytical Technology (PAT) Tool

Yan, B., Zhejiang University
Chen, T.,
Qu, H., Zhejiang University
Originally presented on: 11/6/2013 9:20:00 - 9:45:00

The concept of quality by design (QbD)[1] is now broadly accepted by the pharmaceutical manufacturing industry and widely applied in the process development of pharmaceuticals. In the QbD implementation, usually a design of experiments (DoE) is conducted to build the relationship between material attributes, process parameters and quality attributes. However, it is time consuming and cost intensive to conduct the DoE experiments and then measure the quality attributes of the products or intermediates produced. The additional cost and time have caused some resistance about QbD implementation.[2] Therefore, methods for rapid process development are in urgent need. Integrating process analytical technology (PAT)[3]into DoE is one of the methods worthy of consideration. To show how PAT can enhance rapid process development, this work introduces direct analysis in real time mass spectrometry (DART-MS) as a PAT tool for the DoE study of the chromatographic process of a botanical drug.

Botanical drugs are widely used in China for the prevention and treatment of diseases and chromatographic process is a key purification operation to improve the safety and efficacy of botanical drugs. The chromatographic process usually includes loading, washing, eluting and regeneration stages. Many process parameters are involved in the process and their impacts should be investigated in the process development, according to the QbD initiative. In studying chromatographic loading, dynamic adsorption breakthrough curve of the target component is commonly measured, which can reflect the adsorption capacity of the chromatography column.[4, 5]And the impacts of process parameters on the breakthrough curves are investigated to optimize the adsorption capacity. Traditionally, in the research, many effluent samples at different loading time are collected and analyzed by high performance liquid chromatography (HPLC), which is time and labor intensive. In addition, a proper sampling frequency is difficult to determine before the research. A high frequency leads to increased time of HPLC analysis and a too low frequency may deter the successful plotting of breakthrough curves. Besides, since the breakthrough points are not well recognized during the loading experiments, the material is usually loaded excessively to ensure that the breakthrough points have been reached, which results in additional cost of the material. However, when a PAT tool has been introduced into the research, these issues can be largely addressed.

DART is a novel ambient ionization source that allows direct sampling and ionization in the open air[6]. In the analysis, compounds in the sample are ionized by the electrically discharged gas produced by the DART ion source and then lead to the MS analyzer. In the past few years, DART-MS has been applied successfully in analytical chemistry.[7-9]The sample-preparation-free and noncontact analysis features make DART-MS a promising PAT tool.

In this work, DART-MS is used as a PAT tool for studying chromatographic loading. In the research, a device for the online DART-MS analysis is set up and coupled to the chromatographic process to collect the MS spectra of the effluent. The characteristic ions are selected for determining the target compounds and the breakthrough curves are measured online. The impacts of two parameters (i.e. the concentrations of the target compounds in the material and the flow rate) on the adsorption capacity are studied by DoE. The results show that the PAT tool can be successfully applied to obtain the breakthrough curve, which allows us to get rid of the time consuming offline analysis. A mathematical model between the adsorption capacity and the two parameters studied are established and the adsorption capacity is optimized based on the model. This work has shown the significance of integrating PAT into QbD implementation for rapid process development.

References

[1] International Conference on Harmonisation. "ICH harmonised tripartite guideline: Pharmaceutical development Q8(R2)". 2009. http://www.ich.org/fileadmin/Public_Web_Site/ICH_Products/Guidelines/Quality/Q8_R1/Step4/Q8_R2_Guideline.pdf[accessed 2012-04-28].

[2] Savic, I. M., Marinkovic, V. D., Tasic, L., Krajnovic, D., *Accreditation and Quality Assurance* 2012, 17, 627-633.

[3] FDA. "Guidance for industry, PAT—a framework for innovative pharmaceutical development, manufacturing and quality assurance". 2004. http://www.fda.gov/downloads/Drugs/GuidanceComplianceRegulatoryInformation/Guidances/ucm070305.pdf[accessed Aug 27 2012].

[4] Li, J., Chase, H. A., *J. Chromatogr. A* 2009, 1216, 8759-8770.

[5] Perez-Almodovar, E. X., Carta, G., *J. Chromatogr. A* 2009, 1216, 8339-8347.

[6] Cody, R. B., Laramee, J. A., Durst, H. D., *Anal. Chem.* 2005, 77, 2297-2302.

[7] Chernetsova, E. S., Morlock, G. E., *Mass Spectrom. Rev.* 2011, 30, 875-883.

[8] Hajslova, J., Cajka, T., Vaclavik, L., *Trac-Trends in Analytical Chemistry* 2011, 30, 204-218.

[9] Zeng, S. S., Wang, L., Chen, T., Wang, Y. F., Mo, H. B., Qu, H. B., *Anal. Chim. Acta* 2012, 733, 38-47.

(404d) Measurement Of The Effect Of Total Shear and Compaction Force On Tablet Properties Using Terahertz Pulsed Spectroscopy: Towards The Prediction Of Dissolution Rate

Pawar, P., Rutgers University
Sullivan, M., ADVANTEST AMERICA INC.
Heaps, D., ADVANTEST AMERICA INC.
King, E., ADVANTEST AMERICA INC.
Wang, Y., Rutgers University
Dendamrongvit, W., Rutgers University
Cuitino, A., Rutgers University
Muzzio, F. J., Rutgers, The State University of New Jersey
Originally presented on: 11/6/2013 9:45:00 - 10:10:00

The purpose of this study was to gain an understanding of the effects of process parameters on tablet properties using terahertz pulsed spectroscopy. This was to determine the feasibility for predicting tablet dissolution rates using parameters obtained by transmission terahertz pulsed spectroscopy. Terahertz Spectroscopy was able to detect the effect of both shear and compaction force on tablet properties.

A pre-blend consisting of microcrystalline cellulose, lactose monohydrate, acetaminophen and magnesium stearate was prepared in a V-blender and subdivided into three fractions. Two of the fractions were subsequently sheared for 160 and 640 minutes, respectively, and the third fraction received no additional shearing. Each of the three blend fractions was compressed into tablets using nominal compression forces of 8, 12 and 16 kN. Tablets from each treatment were analyzed non-destructively using an Advantest TAS7500SP pulsed terahertz spectrometer.

Two important terahertz parameters, refractive index (which is a measure of tablet porosity) and peak absorbance (which is proportional to the blend component concentrations and crystallinity) yielded complementary information. The refractive index of the tablets showed a high degree of correlation with compression force, but no distinct trend with shear time. The intensity of the crystalline lactose monohydrate peak in the terahertz absorbance spectra showed a decreasing trend with shear time and minimal influence due to compaction force.

(404e) Implementation of Advanced Hybrid MPC-PID Control System Into a Continuous Pharmaceutical Tablet Manufacturing Pilot-Plant

Singh, R., Rutgers University
Sahay, A., Rutgers University
Brodbeck, P.,
Ierapetritou, M., Rutgers University
Ramachandran, R., Rutgers University
Originally presented on: 11/6/2013 10:10:00 - 10:35:00

There is a need in pharmaceutical industry for the introduction of innovative manufacturing strategies based on continuous processing that is integrated with online monitoring tools, coupled with advanced control systems. This will enable the transition towards a desirable Quality by Design (QbD), rather than Quality by Testing (QbT), based manufacturing of the next generation of pharmaceutical products. This is translated to reduced time, and resources for product development while satisfying the strict regulatory expectations, flexible market demands, operational complexities and economic limitations. Model predictive control (MPC) has proven to be an effective control strategy over the last twenty years and has been widely used in process industry such as oil refining and chemical engineering. In previous work, through a simulation study we have demonstrated that MPC can improve the closed-loop performance of continuous tablet manufacturing process [1, 2]. The implementation of the advanced control system to the pharmaceutical manufacturing plant is still a challenging task because of the different level of complexities involved (e.g. solid handling, irregular flow, sensing challenges, integration of control hardware and software).

In this work, a hybrid MPC-PID control system has been implemented into a direct compaction continuous tablet manufacturing pilot-plant. The continuous feeders and blender are considered here as a demonstrative example. A PAT system is used to read the NIR spectral data at blender outlet and communicate it to the MVA (Multi-Variable Analysis) model performing PCA/PLS to provide the API concentration and RSD (relative standard deviation) value. These critical quality attributes (CQA's) are used as inputs to the MPC in the Process Control System. The MPC uses the two CQA inputs to drive the feed ratio and the blender speed. The MPC output (feed ratio) gives the feeders flow rate set points which are tracked by slave PID controllers. The implemented control scheme utilizes synTQ to integrate DeltaV with an NIR analyzer and MVA model. Main challenges were the integration of multiple hardware/software platforms including Process Control System, NIR analyzer, Multi-Variable Analysis (MVA) software, Process Modeling Software, and Bus Technology. Results demonstrate enhanced performance of critical quality attributes (CQAs) under closed-loop control compared to open-loop operation, thus illustrating the potential of hybrid MPC-PID control system in improving pharmaceutical manufacturing operations. The objective of this presentation is two-fold: first to highlight the design and implementation of hybrid MPC-PID control system for a continuous tablet manufacturing process, and second to demonstrate the performance of the implemented control system.

References

1. Singh, R., Ierapetritou, M., Ramachandran, R. (2013). European Journal of Pharmaceutics and Biopharmaceutics, http://dx.doi.org/10.1016/j.ejpb.2013.02.019.

2. Singh, R., Ierapetritou, M., Ramachandran, R. (2012). International Journal of Pharmaceutics, 438 (1-2), 307-326.

(404f) Real Time Process Management in a Continuous Pharmaceutical Manufacturing Process

Gupta, A., Purdue University
Austin, J. S. III, Purdue University
Nagy, Z. K., Purdue University
Reklaitis, G. V., Purdue University
Originally presented on: 11/6/2013 10:35:00 - 11:00:00

The pharmaceutical industry has traditionally been batch-manufacturing industry but with the advent of Process Analytical Technology (PAT) by FDA, the industry has moved towards continuous manufacturing. The FDA is currently in process of implementing Quality by Design (QbD). Successful implementation of QbD concepts requires understanding of variability in raw materials and the relationship between a process and product's critical quality attributes (CQAs). Under the concept of QbD, when designing and developing a product there is a need to define desired product performance and identify CQAs for the process. The final product attributes could be set on the basis of this information. Therefore, there is a need to understand the impact of raw material and process parameters on the CQAs and identification and control of sources of variability. A process can be continuously monitored and product quality can be ensured, once all these information are acquired.

In this work, a complete plant-wide control framework has been developed for the integrated continuous pharmaceutical tablet manufacturing process. The framework identifies process and product CQAs and defines a set of allowable range of values for each CQAs based on the design space, which is created either using process model or empirically through experiments. Various sensing scheme has been developed to measure different CQAs online that in turn help in real time monitoring of the process. NIR spectroscopy along with Microwave sensing is used to monitor content uniformity of the blend and ribbon density and moisture content of the ribbons. A partial least square model is developed for the measurement of these properties through NIR and a resonance model is developed for microwave sensor. Online measurement of particle size distribution is done through the use of Eyecon imaging technique.

The framework is applied on pilot scale continuous pharmaceutical tablet manufacturing plant consisting of feeders, blenders, roller compacter, granulator and tablet-press. CQA variables were continuously monitored and the framework was able to detect and diagnose various faults, including those involving a single process unit, multiple units as well as controller faults. The framework was able to prevent the process to move into Emergency Shut Down (ESD) and was able to bring it back to Normal Operating Condition (NOC) through the implementation of mitigation strategy.

References:

A. Gupta, A. Giridhar, G.V. Reklaitis and V. Venkatasubramanian, *Intelligent Alarm Systems applied to*

Continuous Pharmaceutical Manufacturing, ESCAPE 23 proceedings, Lappeenranta, Finland, June 2013

A. Gupta, A. Giridhar, V. Venkatasubramanian and G.V. Reklaitis, *Intelligent Alarm Systems applied to Continuous Pharmaceutical Manufacturing – an integrated approach*, Industrial and Engineering Chemistry Research, *DOI:* 10.1021/ie3035042, March 2013

J. Austin, A. Gupta, R. McDonnell, G. V. Reklaitis, M. T. Harris, *The Use of Near Infrared and Microwave Resonance Sensing to Monitor a Continuous Roller Compaction Process*, Journal of Pharmaceutical Sciences, *DOI*: 10.1002/jps.23536, March 2013

(406a) Rapid Process Development of a Pharmaceutical Intermediate for a Late-Stage Compound

Chockla, A. M., The University of Texas at Austin
Benkovics, T.,
Remy, B., Bristol-Myers Squibb Company
Sfouggatakis, C., Bristol-Myers Squibb Company
Originally presented on: 11/6/2013 8:30:00 - 8:55:00

During early stages in development, fit-for-purpose processes are often scaled up rapidly to meet aggressive timelines. While these processes exhibit robust performance in the laboratory, scale-up challenges may be encountered that require process modifications for subsequent campaigns. During the first pilot plant execution of a pharmaceutical intermediate process that consisted of 3-step telescope sequence, long cycle times were observed (~15 days per batch). Seventy percent of this cycle time was attributed to the final reaction step and crystallization / isolation. This talk will highlight how changes to the reaction chemistry allowed for a more streamlined work-up procedure and provided conditions that promoted crystal growth during the crystallization. This talk will also show how DoE strategies and data visualization methods can be used to identify key factors affecting product quality, prove processing ranges, and identify optimal operating conditions. These methods afforded rapid process development on a short timeline that–when implemented at the pilot plant scale–resulted in cycle time reduction of 60-70% (~5 days per batch).

(406b) Development of a pH Driven, Reactive Crystallization of a Pharmaceutical Compound

Cosbie, A. S., Amgen
Huggins, S., Amgen
Jason, T., Amgen
Morrison, H., Amgen
Originally presented on: 11/6/2013 8:55:00 - 9:20:00

Purpose: Develop a robust reactive crystallization and drying procedure for the isolation of the hydrate form of a pharmaceutical compound with consistent particle size distribution and other solid state properties to meet the requirements of the formulation process.

Methods: The crystallization is performed by adjusting the pH of an aqueous solution of substrate with seeding. Effects of seed source and seed load on product particle size are discussed in the development of a growth dominant crystallization. The use of buffered systems and modeling of the equilibrium of the system are used to screen reagents/conditions and to improve the robustness of the crystallization with respect to particle size distribution. Propensity for particle breakage and attrition is investigated. Process analytical tools such as FBRM probes, in-situ pH probes, and hyrgometer are utilized to gain a better understanding of the process.

Results: Seed type and seed load were determined to have a strong influence on the product particle size distribution with the growth dominant system. Inconsistencies in pH at the seed point resulted in seeds dissolving in some cases and inconsistent product particle size distributions. The use of an appropriate buffer system widened the seeding window with respect to pH and improved the reproducibility of the product particle size distribution across multiple scales. The particles' propensity for breakage and attrition during filtration and drying were observed.

Conclusions: A growth dominant, pH driven, reactive crystallization was developed and the factors with the greatest influence on the product particle size were determined. The robustness of the crystallization with respect to product particle size was improved through alternative buffer systems and pH monitoring. The use of a hygrometer on the dryer effluent line was utilized to indicate the end of drying.

(406c) Development of Controlled Agglomeration During Vacuum Agitated Filter Drying for An Active Pharmaceutical Ingredient

Staben Wobker, M., GlaxoSmithKline
Ward, D., GlaxoSmithKline
Hamilton, P. R.,
McKeown, R. R.,
Originally presented on: 11/6/2013 9:20:00 - 9:45:00

In the pharmaceutical industry, agitated filter dryers are used for the isolation and drying of active pharmaceutical ingredients (APIs). A typical drying process has two stages: blowdown, during which filtered nitrogen is passed through the wet API cake to achieve a target solvent level, and agitated vacuum drying, during which vacuum is used to pull off solvent while the wet cake is agitated. This agitated drying can impact the effective particle size through attrition, where a reduction in the primary particle size is observed, and agglomeration, where groups of primary particles bind together to form larger ones. Uncontrolled agglomeration, commonly referred to as balling, occurs when agitated drying begins with too high a solvent concentration in the wet cake. Balling is the formation of very large, dense agglomerates with widely varying sizes, the largest of which can damage drying equipment if undetected. Acicular APIs (high-aspect ratio, needle-like particles) often encounter uncontrolled agglomeration during drying. Developing a better understanding of the agglomeration process can lead to improved control to avoid balling and yield agglomerates with improved flow properties as compared to unagglomerated acicular APIs.

In this study, an innovative drying process was developed to increase the bulk density and flow properties of acicular needles via a controlled agglomeration process. Initial scoping studies were performed using powder rheometry to identify appropriate solvent levels for the initiation and control of agglomeration. Drying experiments were carried out at 1 L, 2 L, and 50 L scale using API that was recrystallized and wet-milled with a rotor-stator mill. The wet-milled API batches were agitated at various "go-drying" loss-on-drying levels identified from the powder rheometry experiments. The knowledge gained from this study was successfully used to control agglomeration of primary particles upon scale up to pilot plant scale.

(406d) Perfusion Cultivation Processes for Monoclonal Antibody Production in Animal Cell Culture Technology: Cell Separation, Critical Challenges and Solutions

Ahmed Elsayed, E., King Saud University, Faculty of Science
A. Wadaan, M., King Saud University, Faculty of Science
Originally presented on: 11/6/2013 9:45:00 - 10:10:00

Mammalian cell cultures are the major source for a large number of very high-value recombinant therapeutic proteins, monoclonal antibodies (MAbs), viral vaccines and hormones. By 2014, it is predicted that five of the top 10 products will be antibodies (Avastin, Humira, Rituxan, Herceptin, and Remicade), along with two other recombinant protein products (Enbrel and Lantus). Furthermore about half of new therapeutic products under development are antibodies. Globally, monoclonal antibodies were expected to have sales exceeding $41 billion in the year 2010. However, efficient and reliable production technologies are required to cope with the tremendous demand for these products. Generally, large-scale commercial mammalian cell cultivations are performed in batch, fed-batch and perfusion processes. Perfusion systems involve cell retention devices, and they have the advantage of continuously removing the product and the exhausted medium, as well as the simultaneous withdrawal of the by-products, specifically toxic metabolites, which largely affect the specific growth rate. Therefore, cells cultured under perfusion culture mode can be maintained at optimal growth conditions. Accordingly, growth inhibition as well as problems with product stability can be reduced.

This presentation will give a short idea about the latest developments achieved in the field of industrial production of monoclonal antibodies with a special attention to the different separation devices used in perfusion cultivations, focusing on one of these systems as a case study.

(406e) Optimal Design of Antibody Downstream Process of Biopharmaceutical Manufacturing

Liu, S., UCL (University College London)
Simaria, A. S., University College London
Farid, S. S., University College London
Papageorgiou, L. G., UCL (University College London)
Originally presented on: 11/6/2013 10:10:00 - 10:35:00

With the increasing market potential of therapeutic monoclonal antibodies (mAbs), it has become critical to rapidly identify the most cost-effective purification processes that can handle increasing upstream productivities in a timely manner and overcome existing purification bottlenecks. Chromatography operations are identified as critical steps in a mAb purification process and can represent a significant proportion of the purification material costs, particularly due to the use of expensive affinity matrices as well as the large amount of buffer reagents required. A key strategic decision relates to the sequence of chromatography steps, which affect the not only the investment and operating cost for the chromatography steps, but also the overall downstream process. Another key decision is the chromatography sizing strategies and this is complicated by the dependencies between steps. There are a large number of possible permutations and trade-offs related to the packed-bed chromatography operations, such as opting for a smaller column run for several cycles to reduce resin costs versus a large column run for fewer cycles to save time.

The aim of this work is to develop an optimisation-based business decision-support framework for the optimal design of chromatography purification steps in mAb manufacture. In this work, a typical mAb purification process with a sequence of three packed-bed chromatography steps was considered. Firstly, a mixed integer nonlinear programming (MINLP) model was developed to determine the optimal chromatography sequencing decisions (resin at each chromatography step) and chromatography column sizing decisions (number of columns, column diameter, column bed height, number of cycles at each chromatography step), so as to minimise the cost of goods per gram (COG/g) of the whole mAb manufacturing process. Due to the computational difficulties to obtain the global optimum of the nonlinear optimisation model, the proposed MINLP model was reformulated as a mixed integer linear fractional programming (MILFP) model using exact linearisation techniques by introducing a number of auxiliary variables and constraints, which was solved by an adopted literature solution approach. An industrially-relevant example with different configurations of upstream and downstream trains was investigated, and the computational results proved the applicability of the proposed models and approaches.

(406f) An Integrated PAT Approach for a Lab Scale Bioreactor IgG3 Cell Culture Process Real Time Monitoring: A Feasibility Study

Wu, H., FDA

Originally presented on: 11/6/2013 10:35:00 - 11:00:00

A model mAb IgG3 cell culture process real time monitoring scheme was reported in this work, to explore the possibility of combining real time spectroscopic techniques and off line characterization tools for a bioreactor cell culture process understanding. Both inline NIR and FTIR Spectroscopy were used to monitor various stages of a model mAb IgG3 cell culture process in real time. In-process nutrient composition, Protein A elution profile, and cell size/morphology were determined offline by Nova 100 Bioanalyzer, Protein A Chromatography, and a digitalized light microscopy, respectively. Without inoculating the bioreactor, a two factor full factorial design (agitation rate with 9 levels, sparge rate with 5 levels) were executed in a 7.5L bioreactor to examine if they impact the NIR real time monitoring. The time series of process NIR spectra was subjected to Principal Component Analysis (PCA). A process trajectory was then constructed based on the PCA results. Three-dimension spectral maps of time-wave number-(NIR or FTIR) spectral intensity were constructed. The process trajectory, 3-D spectral maps, and other process and product characterization results were compared to identify significant process events and elucidate process progression dynamics. It was found that within the scope of DOE, the NIR absorbance of the media was insensitive to either agitation rate or sparge rate. However, the pH adjustments during the bioreactor media equilibration process were characterized by a signal singular point and can be accurately tracked by process trajectory. Applying PCA to the cell culture process NIR clearly demonstrated two distinguishable clusters in the sample map, which correlates well with the Protein A elution profile. The 3-D NIR map demonstrated 3 distinguishable process periods: initial decline, stationary, and growth period. The time series of FTIR spectra along with in-process sample microscopy characterization provided further details regarding some characteristics of those periods. Thus, it demonstrated that combining real time process monitoring and bioreactor in-process sample characterization can help to illustrate process progress and map significant process events.

(465a) Continuous Operations in Pharmaceutical Solid Dosage Form Manufacturing: Dry Granulation Case Study

El Hagrasy, A., Purdue University
Giridhar, A. V., Purdue University
Muliadi, A., Purdue University
Austin, J. S. III, Purdue University
Gupta, A., Purdue University
Louvier, M., Purdue University
Patterson, T. J., Purdue University
Geldenhuis, P., Purdue University
Harris, M. T., Purdue University
Wassgren, C. R., Purdue University
Nagy, Z. K., Purdue University
Reklaitis, G. V., Purdue University
Litster, J. D., Purdue University
Originally presented on: 11/6/2013 12:30:00 - 12:55:00

In recent years, pharmaceutical manufacturing has experienced pressure to change in three interrelated directions. Due to changing market conditions, there is a desire to shift to real-time release (RTR) of drug products to the market soon after manufacturing. The paradigm shift towards RTR requires the implementation of Quality-by-Design (QbD), to obtain a fundamental understanding of the effect of formulation and process variables on the drug product quality attributes, as well as the scientific basis for developing control strategies for the process. The use of process analyzers is another critical component of the QbD framework , for close monitoring of critical process and product parameters and providing the necessary information for use in process control.

In this work, we present a case study in building and operating a continuous dry granulation manufacturing line for production of pharmaceutical tablets at Purdue University as part of the Engineering Research Center for Structured Organic Particulate Systems. We describe the individual powder-handling unit operations: feeding powders, continuous blending, roller compaction, ribbon milling, granule lubrication, tableting. We describe the process data collected using online sensors: NIR, microwave, and X-ray instruments, developed in-house or purchased off the shelf. We describe the automation and process management system: standard DCS features in DeltaV, and custom code developed in-house. We describe our operational procedures developed from both theory and experience. Finally we describe research activities that have been supported by the dry granulation test bed, and how the collected process information is used to improve the manufacturing process.

(465b) Application of An Advanced Process Controller to a Continuous Mixing, Direct Compression Process

Blackwood, D. O., Pfizer Worldwide Research and Development
Mack, J.,
Garcia-Munoz, S., Pfizer Worldwide R&D
Liu, Y. A., Pfizer Worldwide Research and Development
Muteki, K., Pfizer Inc
Originally presented on: 11/6/2013 12:55:00 - 13:20:00

A Model Predictive Controller (MPC) has been applied to a pilot-scale continuous mixing, direct compression manufacturing process. This process consists of three gravimetric feeders, an in-line powder mixing device and a rotary tablet press. A near-infrared (NIR) probe was inserted into the feed frame of the rotary tablet press to continuously monitor blend potency immediately prior to tablet compression. A series of process response tests were applied to the system. Process predictors, such as impeller rotational speed, excipient and API mass feed rate set-points, were varied. The corresponding quality attribute responses, such as blend potency by NIR, mixer hold up mass, and tablet press hopper level were observed. Using this dataset, a dynamic process model has been developed. This model has been used in a Model Predictive Controller to provide multi-variable control of the blend potency prediction by NIR, hold-up mass and total mass flow rate. The prototype predictive controller was challenged by changes to either the total mass flow rate through the system or set point changes to the target blend concentration levels.

(465c) Continuous Manufacturing Traceability for Pharmaceutical Products

Engisch, W. E. Jr., Rutgers University
Muzzio, F. J., Rutgers, The State University of New Jersey
Originally presented on: 11/6/2013 13:20:00 - 13:45:00

Continuous manufacturing has been receiving lots of attention in the pharmaceutical industry with the often repeated obstacle being batch traceability. By definition and inherent nature of a continuous system there is no batch to validate, so redefining traceability and validation techniques in an industry that is dominated by batch manufacturing diction is a necessity.

This work focuses on this solving this problem through the development of method for traceability that is applied to a direct compaction pharmaceutical line. The generalized method that is used can also be easily adapted to the various other continuous manufacturing configurations that may be used in pharmaceutical formulation processing.

Through the use of measured residence time distribution (RTD) of the various continuous processing equipment it is possible to trace any process upsets or disturbances through the entire manufacturing system. Furthermore, the development of proper sampling and analytical techniques are presented and demonstrated with experimental support which can serve to detect and correct any significant deviations and thus mitigating risks.

(465d) Optimized Startup and Shutdown Strategies for Continuous Pharmaceutical Manufacturing

Louvier, M., Purdue University
Giridhar, A. V., Purdue University
Reklaitis, G. V., Purdue University
Nagy, Z. K., Purdue University
Originally presented on: 11/6/2013 13:45:00 - 14:10:00

Continuous manufacturing is a developing area in the pharmaceutical industry with increasing efforts focused on transitioning from the batch to the continuous mode both for drug substance and drug product manufacture. Some of the key requirements for this transition are improved process models, robust on-line sensing technologies, as well as effective supervisory and regulatory control strategies. This research focuses on one aspect of process control, dynamic startup and shutdown control with application to continuous tablet manufacture using a dry granulation process. Being able to optimally control dynamic states of the continuous process is essential for ensuring efficient use of materials, especially the drug substance.

Startup and shutdown control for the process has two overall goals: to minimize the amount of wasted material during the transient phase, and to limit the magnitude of any step changes made, which reduces the strain on equipment and controllers. To achieve these goals, an optimal control strategy is developed, using model-based dynamic optimization techniques to choose optimal setpoints.

The dry granulation line consists of two feeder units, one for Active Pharmaceutical Ingredient (API) and one for excipient, a blender, a roller compactor with attached granulator, a second feeder and blender for adding Magnesium Stearate (MgSt) lubricant, and a tablet press. Instrumentation on the pilot scale equipment is used to both assess the fidelity of the process models and the efficacy of the optimized strategy itself. Each unit op has an associated process model, developed with gPROMS modeling software. These models are used with gPROMS' Optimization package to determine setpoints in the optimal strategies.

The optimal startup strategy begins by determining the optimal final state. Steady state optimization is performed on the process models, maximizing production rate while keeping key process parameters, such as ribbon density, blend composition, and tablet weight, at desired specifications. Dynamic optimization is then performed, which minimizes the mass of "off-spec" material generated during startup. Additionally, constraints are placed to limit the difference between successive set points. The system is brought online in three phases. First, the first two feeders and blenders are turned on and allowed to fill the hopper on the roller compactor. Next the roller compactor is brought online with a series of step changes from the minimum operating speed to the final speed determined. Intermediate set points are determined from the optimization. The granulator and lubricant feeder are also turned on at this point. Finally the tablet press is activated after its hopper is sufficiently full.

The optimal shutdown strategy operates in a similar way. First, the feeders are shut off to prevent more powder from entering the system. While the blender is emptying, the roller compactor is slowed down to its minimum operating speed, and stays at that point until its hopper is empty. The tablet press lowers its production rate to match that of the roller compactor, then shuts down after its hopper is empty. All intermediate setpoints are determined from the dynamic optimization.

These techniques can be compared with other strategies, including a base case where the unit ops are brought online sequentially, to determine their overall effectiveness. This can be done by performing the optimal strategies on the testbed equipment and measuring the amount of offspec material generated for each case.

(465e) Economic Analysis of Batch Vs. Continuous Processing for a Direct Compression Product

Meyer, R. F., Merck & Co. Inc.
Feng, R., Merck and Co. Inc.
Krone-Speck, L., Merck Manufacturing Financial Services
MacConnell, C.,
Reynolds, S. D., Merck and Co., Inc.
Velez, J.,
Originally presented on: 11/6/2013 14:10:00 - 14:35:00

Economic Analysis of Batch vs. Continuous Processing for a Direct Compression Product

Ronnie Feng, Leslie Krone-Speck, Cat MacConnell, Robert Meyer, Scott Reynolds and Jim Velez

Cost pressures in the pharmaceutical industry continue to prompt the search for cost-reducing innovations and new ways of doing business.Â Continuous processing of drug products, where goods such as tablets are manufactured without interruption for arbitrary lengths of time, has been widely cited in the general press as a significant opportunity to reduce manufacturing costs.Â Upon initial examination, continuous processes do appear to provide advantages in comparison to the more common batch processes, most notably through reduction of capital and labor costs and the improvement of production yields.Â The prevalence of continuous processes in lower margin industries such as foods and commodity chemicals further supports the premise that cost advantages associated with continuous processing can be realized in the pharmaceutical industry.Â However, literature and lay press communications are markedly devoid of detailed descriptions of the specific characteristics that lead to quantitative reduction of cost in a pharmaceutical setting, with a notable exception being the 2011 study of S.D. Schaber et al. (Ind. Eng. Chem. Res., 2011, 50 (17), pp 10083?10092).Â

In this study, we describe an assessment of a single new high volume drug product that will be produced via a simple direct compression processing route, encompassing raw material receipt and dispensing, blending, and tablet compression operations.Â Unlike the study of Schaber et al., we assume that manufacture of the active pharmaceutical ingredient (API), drug product manufacture, and packaging steps occur at independent sites. Â We seek to answer the question of whether continuous processing has a cost advantage over other modern, lean batch manufacturing methodologies by comparing the net present cost of manufacturing the drug product using both batch and continuous processing routes.Â Â Â In the case of the batch process (BP), starting materials are received and dispensed to a single batch tumble blender and following blending, this blending tote is docked above a high speed rotary tablet press.Â For the continuous process (CP), starting materials are received and dispensed to individual containers, the containers are docked above gravimetric feeders which supply a pair of continuous blenders, and the continuous blenders subsequently supply a high speed rotary tablet press.Â In both cases, we assume that process analytical technologies (PAT) are used for measurement of blend and tablet uniformity, and thus both the BP and CP utilize a real time release (RTR) testing strategy.Â

To perform a balanced analysis, we assume that the product would require a new facility to be built independent of whether a BP or CP is chosen. Our results indicate that although the more automated CP would require a higher upfront capital investment, the BP would require additional capital investment later in the product life cycle, to increase capacity as product sales volume grows.Â Furthermore, the CP facility size would be marginally smaller than a facility utilizing BP, thus decreasing facility operating costs.Â Although long-term operational data from continuous pharmaceutical processes are not available, we estimate the specific labor for CP and find that it requires less labor to operate, Â and has higher uptime, relative to the BP. Â Elimination of startup and shut down steps for BP results in higher yields for CP, and this yield improvement is quantitatively estimated by analysis of typical, lean BP manufacturing operations.Â Even though RTR testing is employed in both cases, the cost to release each lot of material is still considered, and the larger lot sizes produced by the more automated CP are estimated to require fewer man hours to release.Â Finally, shorter production cycles and less work-in-progress would result in inventory cost savings for the CP relative to the BP.

Summation of the various cost differences shows that a CP would be financially favorable over a BP, a result which is shown to be true under a variety of scenarios tested in a sensitivity analysis.Â As expected, the magnitude of the savings is dependent on the production volume of the product being manufactured, with higher savings realized at higher production volumes.Â However, the magnitude of the calculated cost savings is smaller than what has been reported elsewhere. This result may stem from assumptions used in other studies where the batch production models have not been ?leaned out? to the maximum extent possible. Ultimately, it can be concluded that a CP can be a lower cost option for drug products which require capital investment in a new facility, but this lower cost must be balanced against the uncertainty and risk which come with new operational methods. A key finding in this study is that the benefit to risk ratio may be significantly smaller than is often noted in more informal reports on continuous processing of pharmaceuticals.

(465f) Applying Economic Analysis in the Evaluation of Batch Versus Continuous Manufacturing for Oral Solid Dosage Forms

Fraser, S. D., Research Center Pharmaceutical Engineering (RCPE)
Dujmovic, D., RCPE GmbH
Khinast, J. G., Graz University of Technology
Originally presented on: 11/6/2013 14:35:00 - 15:00:00

Up to now, the introduction of continuous manufacturing in the pharmaceutical industry has been discussed with a strong focus on technological aspects and advantages. This is due to the significant technological and scientific challenges faced in moving from isolated end-point-driven unit operations, and their straight-forward batch definition, to semi-continuous or even plug flow approaches. In addition, the commercial availability of integrated continuous manufacturing lines is currently still limited, as is the operational experience with these lines. These technological and scientific challenges are currently addressed by various research groups in academia, as well as by the pharmaceutical and equipment engineering industry.

Looking just a few years into the future, different continuous manufacturing lines will be readily available from pharmaceutical equipment vendors. Pharmaceutical companies, on the other hand, will have gained operational experience in commercial manufacturing operations with these lines. The question whether or not to apply a continuous or a batch-based manufacturing approach will then be answered by the actual process requirements, as well as the associated economic dimension in terms of cost of goods manufactured. The selection of the technological manufacturing approach will then be less of a choice between a more traditional (batch-based) and a more innovative (continuous) approach, but more of a holistic techno-economic optimization exercise.

Based on the experience of previous and ongoing continuous manufacturing development projects, a general methodology of how to include an economic analysis into a pharmaceutical process selection is presented. Possibilities and limitations of the methodology are introduced with a conceptual process evaluation example dealing with secondary manufacturing of an oral solid dosage form. Implications on the relevance of continuous processes in future pharmaceutical manufacturing operations are briefly discussed.

(466a) A Continuous Flow Grignard At Scale

Nwosu, S. O., Eli Lilly and Company
Johnson, M. D., Eli Lilly and Company
Adler, J. J., D&M Continuous Solutions
Schafer, J. P., D&M Continuous Solutions
Braden, T. M., Eli Lilly and Company
Kerr, M. S., Eli Lilly and Company
Seibert, K. D., Eli Lilly and Company
Kopach, M., Eli Lilly and Co.
Originally presented on: 11/6/2013 12:30:00 - 12:55:00

Grignard reactions are common to pharmaceutical, fine and specialty chemical industries. These reactions are typically very exothermic with large adiabatic temperature rises, and are often water sensitive. Maintaining an inert atmosphere condition thus becomes critical throughout production, storage and use in order to maximize safety and yield.

To improve the overall safety and control of a Grignard reaction at production scale, a CSTR (Continuous Stirred Tank Reactor) was installed to improve inertion capability, and exotherm control, while maintaining cycle-time with a significantly reduced reactor size. This route of operation has the added advantage of minimizing excess magnesium needed, and allowing for magnesium run out at the end of the cycle, thus minimizing the overall excess magnesium metal that will need to be disposed of at the end of the process.

Results will be presented to demonstrate the technical challenges and successful scale-up of the Grignard step from the laboratory, through the pilot plant, and into full scale production.

(466b) Packed-Bed Hydrogenolysis: A Multi-Scale Study

Zaborenko, N., Eli Lilly and Company
Linder, R. J.,
Campbell, B. M.,
Originally presented on: 11/6/2013 12:55:00 - 13:20:00

In recent years, Eli Lilly chemical process development has devoted a great deal of focus on continuous-flow reaction and workup platforms for process intensification, increased safety, expanded range of reaction conditions, and improved plant space utilization. This talk will disclose our design and development of a continuous-flow packed catalyst bed system as applied to a two-step hydrogenolysis. The deprotection and ring-opening of an *N*-protected isoxazolidine substrate have been accomplished by use of a Pd/C catalyst in an aqueous alcohol solvent with hydrochloric acid, affording a hydrochloride salt solution. Numerous flow reaction studies have been performed to characterize the reactors and catalyst used for this purpose. The continuous-flow process resulted in a 60× reduction of residence time, improved impurity control strategy, and increased safety as compared to the batch process developed for this reaction. The process applied several in-line HPLC strategies for process analytical technology, providing real-time actionable process information and enabling rapid response to process departures and to changes in catalyst activity. Reaction scales from 10 to 600 g/hour of substrate and up to 10 kg of isolated product will be presented, demonstrating the utility and advantages of the non-cGMP reactor and equipment configuration versus batch, as well as the ability of continuous-flow technology to enable pilot-plant-scale or even manufacturing-scale production within laboratory-scale footprints.

(466c) Design and Scale-Up of the Continuous Reactive Crystallization of An API

Polster, C. S., Eli Lilly and Company

Originally presented on: 11/6/2013 13:20:00 - 13:45:00

Design and Scale-Up of the Continuous Reactive Crystallization of an API

Christopher S. Polster

Eli Lilly and Company, Small Molecule Design and Development

Abstract:

A continuous reactive crystallization of an active pharmaceutical ingredient (API) was designed to increase process throughput and product consistency. The process uses high shear mixing elbow injectors in order to enable a process to grow crystals in a controlled fashion, allowing for direct control of product particle size and powder physical properties. The work described in this presentation includes the analysis of crystallization kinetics, the use of kinetics to enable modeling for process scale up and optimization as well as the correlation of several measurable characteristics of API which correlate directly to drug product blend flow. Crystallization kinetics are elucidated through the usage of gCRYSTAL software; both growth and nucleation kinetics are modeled. The kinetics obtained are used to compare rates of nucleation, growth and liquid/liquid mixing phenomena, establishing limits on processing parameters. Equipment capable of 10 kilograms of throughput over 72 hours in a 2.5 L crystallizer volume is described.

(466d) Use of Continuous Msmpr Crystallization With Integrated Nanofiltration Membrane Recycle for Enhanced Yield and Purity in API Crystallization

Ferguson, S. T., Massachusetts Institute of Technology
Ortner, F., TU Munich
Quon, J., Massachusetts Institute of Technology
Peeva, L. G., Imperial College London
Trout, B. L., Massachusetts Institute of Technology
Livingston, A., Imperial College London
Myerson, A. S., Massachusetts Institute of Technology
Originally presented on: 11/6/2013 13:45:00 - 14:10:00

Crystallization is the most widely used separation and purification technique in the pharmaceutical industry. Therefore, it is essential that continuous crystallization techniques are available to meet or exceed the purity and yield achievable in current batch crystallization processes if they are to be widely utilized in pharmaceutical production. Continuous tank based MSMPR crystallizers operate at steady state at a fixed position on the phase diagram for a given crystallization. This means that the process can operate at a set of conditions (temperature, solvent composition, supersaturation etc.) that are kinetically favorable for a given crystallization. In cases where these kinetically favorable conditions require a higher liquid phase solute concentration than the alternate batch crystallization (at final isolation conditions), there would be an associated loss in yield. Any such loss in yield can be eliminated by recycling concentrated mother liquor back into the MSMPR. An evaporation based mother liquor recycle strategy enabled improved crystallization performance in terms purity for the API cyclosporin compared with the current commercial batch process. [1]

Evaporation based concentration and recycling of mother liquor recycle has two major limitations that can be mitigated by the use of a nanofiltration membrane module in its place. Firstly, evaporation will change the solvent composition in mixed solvent and solvent anti-solvent systems, thus limiting this techniques applicability. This is not an issue for nanofiltration concentration as it works through selective molecular sieving. In this case, API molecules are preferentially retained by the membrane relative to the smaller solvent molecules. The concentrated retentate stream from the membrane is then recycled back to the crystallizer. In this work a nanofiltration module is successfully used to concentrate the recycle stream for the MSMPR crystallization of an API from a mixed solvent system (THF:Ethanol) using combined cooling anti-solvent (Water) approach to generate supersaturation.

In both the evaporative concentration and the nanofiltration module, the limiting factor on yield is the purification requirements for the API in question. For this reason API containing solution must be purged at a rate that prevents the build up of impurities in the system, thus reducing the process yield. However, nanofiltration can also allow preferential removal and purging of impurities that have a reasonable large deviation in structure or molecular weight compared to a given API. [2] This was the case for the crystallization of the API in this study (MW=373 g/mol) where the most significant impurity had a lower molecular weight of 152 g/mol. Through screening experiments, a membrane that allowed preferential permeation of impurity was found and incorporated into the MSMPR recycle process. This negated the need for the use of a purge stream to reduce impurity concentration, resulting in significantly higher yield and comparable or improved purity than would be attainable in batch or standard MSMPR crystallizations.

[1] Wong, S.Y., Tatusko, A.P., Trout, B.L., Myerson, A.S. (2012). Development of Continuous Crystallization Processes. Using a Single-Stage, Mixed-Suspension, Mixed-Product Removal Crystallizer with Recycle. Cryst. Growth & Des. 12(11), 5701-5707.

[2] Sereewatthanawut, I., Lim ,W., Bhole, Y.S., Ormerod, D., Horvath, A., Boam, A.T., Livingston, A.G. (2010). Demonstratio of Molecular Purification in Polar Aprotic Solvents by Organic Solvent Nanofiltration. Org. Proc. Res & Des. 14, 600-611.

(466e) Model-Predictive Control Strategies for Improved Bioprocess Performance

Glennon, B., UCD
Whelan, J., University College Dublin
Barrett, M., University College Dublin
Originally presented on: 11/6/2013 14:10:00 - 14:35:00

There is an increase in interest within the biopharmaceutical industry to move from a quality-by-inspection to a quality-by-design (QbD) approach towards process design, optimisation and operation. The implementation of this philosophy through the use of a PAT-enabled control strategy for a bioreactor is decribed in this work. In this study, a number of elements were developed and integrated to implement advanced feedback control of substrate concentrations in a fed-batch CHO cell bioprocess. It is desirable to control substrate and by-product concentrations to provide better process performance in terms of cell density, culture longevity and protein quantity and quality. To implement such control, a Raman spectroscopy method for the simultaneous, real-time measurement of cell density, glucose, glutamine, glutamate, lactate and ammonia was developed. Secondly, a first-principle engineering model of the fed-batch process was identified which facilitated process control simulations. The simulations identified model predictive control (MPC) as a promising form of process control for the inherently complex and highly variable nature of bioprocesses due to its ability to reject measurement noise, handle long sample intervals, cope with non-linear processes and operate as a multiple-input-multiple-output (MIMO) control strategy. Finally, MPC control of glucose and glutamine concentrations was successfully implemented on 3 L and 15 L bioreactors. In transitioning from a bolus to continuous feeding strategy, a 30 % increase in cell density was achieved.

(466f) Small Scale Media Optimization for Continuous Culture - Effect On Cellular Metabolism

Karst, D., ETH Zurich
Morbidelli, M., ETH Zurich
Soos, M., ETH Zurich
Originally presented on: 11/6/2013 14:35:00 - 15:00:00

The continuous culture of mammalian cells is designed to produce biotherapeutics at high cell density and viability at steady state condition up to months. Thereby, to mimic the physiological cell environment media formulation is one of the most critical process parameters to tune cellular metabolism optimizing growth, viability and productivity. Both the quantitative composition and concentration are known to affect named culture attributes.

In this study various media compositions including primary carbon sources such as glucose and glutamine, as well as supplemental amino acids, trace elements and vitamins are systematically tested in small spinning tube batch cultures. Preselected conditions are transferred to determine their performance in a perfusion culture set up using tangential flow filtration for cell retention. Changing media composition result in distinct steady states of the cell culture, allowing the identification of nutritional needs when facing increased cell densities. Finally, using the experimental data our goal is to characterize the metabolic state of the cell during optimal culture conditions based on a metabolic network applying Flux Balance Analysis (FBA). Information about the activity of certain metabolic pathways, such as glycolysis, TCA cycle or amino acid metabolism is expected to result in better understanding of the cellular requirements during high density perfusion culture.

Gained knowledge will be used to improve the ability of constructing an enhanced culture environment for mammalian cells and will help in tuning their metabolic state with respect to growth and productivity.

(503a) Selection and Crystallization Development of a Practical Salt Form for a Pharmaceutical Compound

Yang, B. S., Boehringer Ingelheim Pharmaceuticals, Inc.
Li, Z., Boehringer Ingelheim Pharmaceuticals, Inc.
Jiang, M., Massachusetts Institute of Technology
Kim, S., Boehringer Ingelheim Pharmaceuticals, Inc.
Wang, X.,
Yee, N., Boehringer Ingelheim Pharmaceuticals, Inc.
Senanayake, C., Boehringer Ingelheim Pharmaceuticals, Inc.
Originally presented on: 11/6/2013 12:30:00 - 12:50:00

This paper presents a fully integrated solid form screen and crystallization development workflow for the development of an active pharmaceutical ingredient (API) at BI. Salt I of an in-licensed compound was initially used in the Phase I and Phase IIa studies. However, Salt I was later characterized as an ethanolate/hydrate and had several development issues, including the high and variable moisture content and unstable physical/chemical properties under various conditions (heat, humidity, mechanical and light). As a result, a new salt form that can provide better developability was required that triggered extensive salt/polymorph screen and selection activities, in order to select a new salt form to be used for Phase IIb and beyond.

After a thorough solid form screen and selection process, Salt II, with its superior physicochemical properties to the other crystalline salts identified, was chosen as the new development form. In the initial development, a workable process (Process I) was developed in EtOH/Heptane that was able to produce multiple kilo scale of Salt II to supply for the formulation development. The particle size was controlled by dry milling initially. Later, a more comprehensive crystallization development was conducted to address several development challenges, including the purification of the major impurity from the synthesis, hydrolysis of the compound in the crystallization system, color removal and powder property control, etc. A new modified process (Process II) was successfully developed, enabled by the selection of appropriate solvent system and design of crystallization condition that can inhibit the hydrolysis of the compound during crystallization. Process II used milled seeds and controlled crystallization to directly produce the desired particle size distribution without milling. Process II was successfully scaled up to multiple 30 kg batches in the Pilot Plant.

(503b) In-Situ Monitoring of Resolution of Chiral Compounds Using Process Analytical Technology

Chen, Z., National Engineering Research Center for Industrial Crystallization Technology
Gao, Y., National Engineering Research Center for Industrial Crystallization Technology, School of Chemical Engineering and Technology, Tianjin University
Wang, Y., National Engineering Research Center for Industrial Crystallization Technology, School of Chemical Engineering and Technology, Tianjin University
Yin, Q., National Engineering Research Center for Industrial Crystallization Technology, School of Chemical Engineering and Technology, Tianjin University
Zhang, M., Tianjin University
Hou, B., Tianjin University
Hao, H., National Engineering Research Center for Industrial Crystallization Technology, School of Chemical Engineering and Technology, Tianjin University
Originally presented on: 11/6/2013 12:50:00 - 13:10:00

In-situ Monitoring of Resolution of Chiral Compounds using Process Analytical Technology

Zhen Chen, Yuan Gao, Yongli Wang, Qiuxiang Yin, Meijing Zhang, Baohong Hou, Hongxun Hao*

National Engineering Research Center for Industrial Crystallization Technology, School of Chemical Engineering and Technology, Tianjin University, Tianjin 300072, P.R. China

* Corresponding author: Tel:+86-22-27405754 Fax: + 86-22-27405754

Email: hongxunhao@tju.edu.cn

Abstract

Since chirality was discovered and defined hundreds of years ago, the resolution and purification of chiral compounds has become one main challenge that scientist have to face. Most physicochemical properties of the enantiomers are nearly the same or pretty close. So it is quite difficult to find an appropriate method to resolve the chiral compounds, especially in industry scale. As one main resolution method of chiral compounds, preferential crystallization has been widely used not only in the labs but also in industries. To develop and design an optimum crystallization processes for chiral compounds, it is crucial to understand the thermodynamic and kinetic mechanism of the processes. Process analytical technologies (PAT) can be used to monitor the preferential crystallization processes in line and are very helpful to further understand the mechanism behind the phenomena. In this paper, several PAT tools were successfully used to in-situ monitor the resolution of L-lactide. Raman Spectroscopy and attenuated total reflectance Fourier transformation infrared (ATR-FTIR) were used to monitor the solid phase and liquid phase concentration respectively. Focused beam reflectance measurement (FBRM) and particle vision measurement (PVM) were used to monitor the particle size and morphology respectively. Some other offline analytical tools, such as X-ray diffraction (XRD), differential scanning calorimetry (DSC), Automatic polarimeter were also used to confirm the online analysis results. Solubility data of L-lactide were also measured experimentally. The effect of operating parameters, such as seeding, cooling profile and the initial concentration, on the preferential crystallization processes was also investigated. It was found that all of these operating parameters can affect the resolution results. The reasons and the mechanism of the influence were discussed in detail. Using these knowledge obtained in this paper, an reliable optimum resolation method with high output capacity and good repeatability for L-lactide were finally developed.

Keywords: Resolution, Chiral compounds, Process analytical technology, Preferential crystallization, L-Lactide

(503c) A Model for Crystal Growth for Solids Crystallizing From Solutions Containing Multiple Conformers and Evidence of Growth Inhibition By the "Wrong Conformer"

Derdour, L., Bristol-Myers Squibb co.
Sivakumar, C., Carnegie Mellon University
Skliar, D., Bristol-Myers Squibb Co.
Pack, S., Bristol-Myers Squibb Co.
Lai, C., Bristol-Myers Squibb Company
Vernille, J., Bristol Myers Squibb
Kiang, S., Bristol-Myers Squibb Company
Originally presented on: 11/6/2013 13:10:00 - 13:30:00

The validity of a model derived for crystal growth rate for solids crystallized from solutions containing multiple conformers is evaluated. A simple and straightforward method for determining crystal growth rates using FBRM and FTIR/ATR is utilized to maximize growth rate data collected from a single experiment. Crystal growth rate data obtained experimentally were in good agreement with model predictions. In particular, the data indicated that when the approach of the right conformer (Derdour et al. Chem Eng Sci., 66(1), pp. 88-102, 2011) applies, crystal growth rate is inversely proportional to crystal size. Further, the results indicate that growth rates determined from supersaturation measurements can be over-estimated because of surface nucleation and/or agglomeration that can result from supersaturation spikes during anti-solvent addition. Lastly, crystal growth data obtained at different temperatures indicated that the slow growth rate observed at low temperatures may be caused in part by surface poisoning by the wrong conformer.

(503d) API Particle Engineering for Drug Substance: Crystallization & Wet Milling to Control API Powder Properties

Schild, R., Bristol-Myers Squibb
Engstrom, J. D., Bristol-Myers Squibb Company
Hubert, M., Bristol-Myers Squibb Company
Jones, J.,
Qian, X.,
Wu, Y., Bristol-Myers Squibb Co.
Leane, M.,
Gamble, J., Bristol-Myers Squibb
Wang, S. S. Y., Bristol-Myers Squibb Company
Originally presented on: 11/6/2013 13:30:00 - 13:50:00

In the pharmaceutical industry, target drug substance powder properties, such as particle size and flow are defined for a given formulation to ensure robust drug product processing. Batch crystallizations are primarily designed to maximize the active pharmaceutical ingredient (API) yield and product purity and frequently produce API's with non-ideal size and morphology. In this investigation, the effect of seeding, supersaturation, and post crystallization wet milling were used to produce API batches with a range of particle sizes and flow properties to define the initial design space of acceptable content uniformity and powder flow. For laboratory and pilot plant scale-up experiments, the particle size was measured real-time during crystallization and wet milling using focused beam reflectance measurement (FBRM) and correlated to spot slurry samples analyzed by laser light scattering as well as image based particle characterization. These results were compared with empirical correlations to account for initial particle size, mill generator configuration, rotor or tip speed, milling residence time, and rotor/stator gap or tooth dimensions to aid successful crystallization and wet milling scale-up.

(503e) Hysteresis in Solid-Liquid Suspensions in Optimax®: Using CFD Simulations and Experiments With FBRM

Vedantam, S., National Chemical Laboratory

Ranade, V. V., National Chemical Laboratory

Originally presented on: 11/6/2013 13:50:00 - 14:10:00

Title: **Hysteresis in solid-liquid suspensions in OPTIMAX®: Using CFD simulations and experiments with FBRM**

*Authors: Vedantam, S., Ranade, V.V.**

Affiliation: Industrial Flow Modeling Group (iFMg), CEPD Division, National Chemical Laboratory (NCL), Pune, INDIA

**corresponding author: vv.ranade@ncl.res.in*

ABSTRACT

Introduction:

Crystallization is the most widely used separation and purification process employed throughout the chemical and pharmaceutical industry for chemical products that are solids at room temperature and pressure. Today, crystallization modeling tools are limited by the availability of methods to reliably predict solubility and super saturation for chemical systems of industrial importance. There is also limited capability for describing nucleation and crystal growth, and mixing and design scale-up for industrial crystallizers. Tools are needed to model and optimize crystallizer process performance with respect to product quality, throughput and efficiency. It is essential to understand and relate key thermodynamic, kinetic and hydrodynamic aspects to crystallizer performance not just in terms of yield but also in terms of product quality (characterized by particle size distribution, morphology, polymorphism and the amount of strain as well as the uptake of solvent or impurities in the crystal lattice). Process Analytical Technology (PAT) tools are also on the rise in these applications and published in the literature (Daneau and Steele., 2005; Samad *et al.*, 2013).

Hydrodynamic modeling, one of the key aspects, is the focus of the work presented in this abstract. In literature, Joshi et al (2011), presented a comprehensive review on CFD modeling studies in stirred tanks, stating that it is worth noting that dense suspensions require a correct modeling of both solid?liquid and particle? particle interactions. However, as an initial step, an automated reactor called OPTIMAX® (launched by Mettler-Toledo Ltd.,) is being used for the study of various aspects of crystallization and the quantitative effect of all key aspects on the process. OPTIMAX® (OM), a synthesis workstation provides a platform for product and process development and is coming into use in bio-pharmaceutical synthesis applications. OM is typically an automated reactor, offering heating electrically and cooling on the basis of Peltier technology. Crystallization is one of the wide hopes of applications to be carried out in this particular workstation. In this particular set-up, OM is designed to accommodate internals such as Focused Beam Reflectance Measurement (FBRM) probe (used for estimating the Chord Length Distributions (CLD) and thus, Crystal Size Distributions (CSD)); Temperature probe (Tr probe), Fourier-Transform Infra Red (FTIR) spectroscopy probe; turbidity probe; pH probe; impeller and baffles as well. However, as for this study, the internals used are FBRM probe; Tr probe and 4-bladed down-pumping PBT as impeller are used. It is understandable that the reactor geometry, operating conditions and the flow characteristics govern the hydrodynamic aspects of the crystallization process thus subsequently also affecting the kinetic aspects. Solid-liquid suspension studies on solids distribution and the effect of operating conditions has been simulated and experimentally verified.

Methodology, flow characterization and mixing time:

Commercial software Ansys Fluent (14.0) has been used for all CFD simulations. Multiple Reference Frame (MRF) approach is used for modeling the impeller. Multiple phases have been simulated using Eulerian approach. 1000 mL reactor volume has been used as the domain for simulations and experiments, with an inner diameter of 0.101 m, and a height of 0.137 m. The impeller used is a 45^0, four-bladed, down pumping PBT. A mesh size of 4, 60,000 cells has been used for the entire domain. Simulations are carried out for impeller speed ranging from 100 to 1000 rpm, and varying solids hold-up of the particles used. The geometry indicating the internals, tank and inner and outer regions of the domain, clearly demarked for MRF approach are shown in the Figure 1. The commonly used validation check points such as Power number (N_P, from turbulent energy dissipation rate and moments), Pumping number (N_Q) are estimated from the CFD predictions, and satisfactory validation has been obtained, which gives a go ahead for further studies on crystallization in this reactor.

Experiments are carried out using FBRM to estimate the distribution of particles, and verify the hysteresis likely to occur with the variation in impeller speed. Neutrally buoyant particles of 350 micron are used as solid particles. Hold-up has been varied from 1 % to 10 % in the reactor. Impeller speeds are varied from 100 rpm to 1000 rpm and the particle size distribution has been tracked using FBRM. Similar conditions have been simulated using CFD.

Conclusions:

Solid-liquid suspension studies in OPTIMAX® are carried out using simulations as well as experiments. The CFD study is validated with the established parameters such as N_P (from estimated turbulent energy dissipation rate and the torque), N_Q from the CFD predictions. Possibility of hysteresis has been verified using FBRM and CFD simulations.

References:

Samad, N.A.F.B.A., Sin, G., Gernaey, K.V., Gani., R. 2013, ?A systematic framework for design of process monitoring and control (PAT) systems for crystallization processes?, Computers & Chemical Engineering, 54, 8-23. http://dx.doi.org/10.1016/j.compchemeng.2013.03.003

Buwa, V. V. and Ranade, V. V., 2003a Characterization of dynamics of gas-liquid flows in rectangular bubble columns, AIChEJ, 50, 2394.

Deneau, E., Steele, G, 2005, An in-line study of oiling out and crystallization, Org. Proc. Res. Dev., 9, 943-950.

Joshi, J.B., Nere, N.K., Rane, C.V., Murthy, B.N., Mathpati, C.S., Patwardhan, A.W., Ranade, V.V., 2011, CFD Simulation of Stirred Tanks: Comparison of turbulence models (Part II: Axial Flow Impellers, Multiple Impellers and Multiphase Dispersions), The Canad. J. of Chem. Eng., 89, 754-816.

Ranade, V. V., Bourne, J. R. and Joshi, J. B., 1991. Fluid mechanics and blending in agitated tanks. Chem. Eng. Sci., 46, 1883.

Figures:

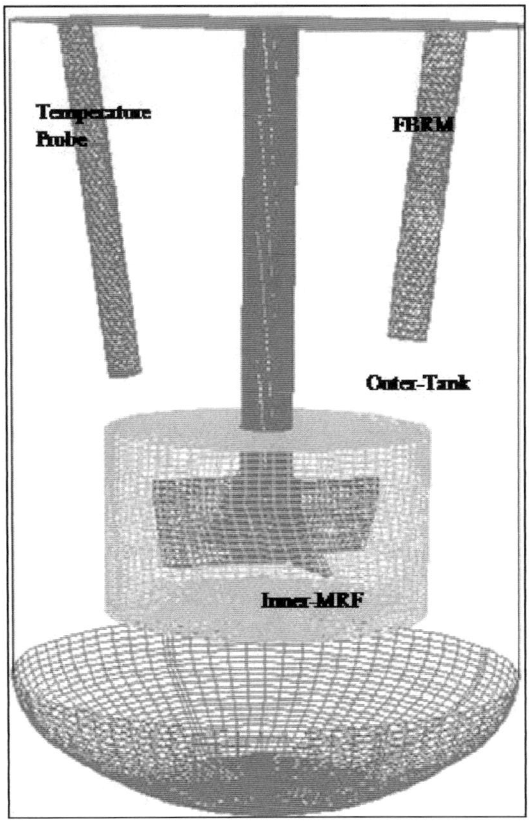

Figure 1. Computational domain showing inner and outer regions of MRF approach

(503f) Development of a Model for Anisotropic Crystal Growth From Solutions

Derdour, L., Bristol-Myers Squibb co.
Chan, E. J.,
Originally presented on: 11/6/2013 14:10:00 - 14:30:00

A model for anisotropic growth was developed. It is based on an isotropic growth model recently developed for crystal growth from solutions containing multiple conformers. The model presented herein implies that growth rates are constrained in one spatial dimension, thus only two linear growth dimensions, namely length and width, are considered. Crystal growth is assumed to occur via step propagation. The model takes into consideration step advance velocity dependence upon supersaturation and solvent/crystal interaction. An approach which uses two solute integration coefficients is introduced that describes crystal growth kinetics in each growth dimension. These coefficients are functions of the linear growth of the faces and are inclusive of solvent/crystal facet interactions. A parameter estimation algorithm to determine these coefficients is described. The algorithm requires experimental data of concentration variation during crystallization and final aspect ratio. The model's predictions of temporal variation of supersaturation and final crystal elongation factor were in agreement with experimental data for a Bristol-Myers Squibb proprietary compound. Finally, model simulations allow for the estimation of integration coefficients for any given aspect ratio.

(503g) A Straightforward Technique to Control the Oxidation of Amorphous Cobalt Nanoparticles During Synthesis Using DMSO

Duggan, J. N., Auburn University
Roberts, C. B., Auburn University
Chaturvedi, A.,
Suzuki, T.,
Originally presented on: 11/6/2013 14:30:00 - 14:50:00

Magnetic nanoparticles have attracted a great deal of interest lately because of their unique magnetic properties and have a wide range of potential applications. For example, nano-scale magnetic particles have enhanced magnetic susceptibility and saturation compared to their bulk-scale counterparts. Cobalt nanoparticles can exhibit a superparamagnetic behavior, which makes these particles particularly useful as contrast agents in Magnetic Resonance Imaging (MRI). However, many of the techniques commonly used to produce well-defined cobalt nanoparticles involve the use of harsh reagents and synthesis conditions, and may require days to produce a single sample of nanoparticles. Furthermore, it is often difficult to produce cobalt nanoparticles because they have a tendency to become overly oxidized, and can subsequently agglomerate and precipitate from the solution.

We have developed a single-step method to synthesize and stabilize cobalt nanoparticles using an FDA-approved, functional solvent, dimethyl sulfoxide (DMSO). Moreover, this research demonstrates that DMSO can be a multi-functional component in nanoparticle synthesis by serving concomitantly as a solvent and as a stabilizing agent that effectively suppresses nanoparticle growth and prevents excessive oxidation of the nanoparticle surface. The size and morphology of the particles were determined using TEM imaging analysis. In addition, we have performed a variety of different characterization techniques to investigate the interaction of the cobalt nanoparticles with the DMSO stabilizing molecules, including FT-IR, XPS, and XRD. Using these techniques, we found that both the sulfur and oxygen functional groups on the DMSO molecules interact with the nanoparticle surfaces via a resonance hybrid structure that stabilizes the nanoparticles. We have also found that these cobalt nanoparticles are amorphous when synthesized at room temperature. Ongoing studies in our lab are focused on subjecting the cobalt nanoparticles to a variety of different temperatures post-synthesis in order to determine the minimum temperature needed to induce oxidation and crystallinity within the particles. In addition, the magnetic behavior of these amorphous cobalt nanoparticles compared to the magnetic behavior of crystalline cobalt and cobalt oxide nanoparticles will be discussed.

(536a) Crystallization of L-Ascorbic Acid in a 18-L Airlift Crystallizer

Lakerveld, R., Delft University of Technology
Van Krochten, J., Delft University of Technology
Anisi, F., Delft University of Technology
Stankiewicz, A. I., Delft University of Technology
Kramer, H. J. M., Delft University of Technology
Originally presented on: 11/6/2013 15:15:00 - 15:40:00

Crystallization is applied extensively in the chemical and pharmaceutical industry to produce and purify solid particles. The performance of the process and final product are closely related to specific properties of the crystals such as purity, crystal size distribution (CSD), and shape. However, the design and control of industrial crystallization processes to manufacture a crystalline product with tailored properties remains a major challenge. Many interacting physical phenomena occur in solution crystallization processes, involving for example primary nucleation, growth and attrition of crystals. In current industrial equipment, those physical phenomena take place in the same environment and are, therefore, strongly entangled. Novel designs are needed to improve control over individual physical phenomena.

Attrition refers to the birth of new crystals caused by collisions of crystals, which is often undesirable as it reduces the crystal size and broadens the CSD. Recently, a new crystallization concept involving an airlift, which eliminates the need for a stirrer, has been tested for crystallization of ammonium sulphate from water[1]. The results demonstrate that attrition can be significantly suppressed compared to crystallization in a conventional stirred crystallizer.

Although these results are very promising for further development, the full potential of the concept remains largely unclear. An airlift crystallizer offers an exciting opportunity to bring the control of crystal product quality to a new level particularly for systems that are very sensitive to the mechanical impact of a stirrer. In particular, pharmaceutical compounds that are mostly produced batch-wise and in stirred-tank crystallizers often suffer from such sensitivity, which limits flexibility to produce a desired CSD due to uncontrolled nucleation. In order to apply an airlift system for crystallization successfully, the process conditions need to be optimized. Especially the control of the flow regime and the design of the sparger are essential for the performance of this new type of crystallizer.

The objective of the present study is to investigate experimentally the performance of an 18-L airlift crystallizer for crystallization of L-ascorbic acid from water in terms of hydrodynamic behavior and final product quality for seeded batch crystallization. First, the hydrodynamic behavior of the system has been investigated as function of the operating conditions and the design of the sparger via several in-line density measurements and in-line imaging. The results demonstrate that favorable hydrodynamic conditions in which the injected nitrogen is only present in the riser of the system can be achieved via a proper design of the sparger for a broad range of operating conditions. Secondly, the attainable product quality in terms of shape and CSD has been investigated via in-line imaging and off-line measurement of the supersaturation profile and CSD. A systematic comparison between the crystallization performance in an 18-L airlift crystallizer and crystallization in a conventional 5-L stirred crystallizer has been conducted. The results demonstrate a significant improvement of crystal shape and reduced nucleation in the airlift crystallizer. Furthermore, a clear distinction between grown seeds and new nuclei in both crystallizers allow us for the first time to quantitatively distinguish various forms of nucleation of the studied model system. Thirdly, the experimental data has been used to develop kinetic expressions for growth and nucleation, which allows for model-based optimization of the system to reveal the full potential of both a conventional crystallizer and airlift crystallizer. Finally, prospects for extending the system to continuous flow-mode will be discussed.

References

[1] A. Soare, R. Lakerveld, J.van Royen, G. Zocchi, A. I. Stankiewicz, H. J. M. Kramer. Minimization of Attrition and Breakage in an Airlift Crystallizer. *Ind.Eng.Chem.Res.* **2012**, 51, 10895-10909.

Acknowledgement

This work was supported by the European Commissions Framework 7 program through the OPTICO consortium

(536b) Mathematical Modeling and Analysis of Cooling Crystallization within Dual-Impinging-Jet Mixers

Jiang, M., Massachusetts Institute of Technology
Braatz, R. D., Massachusetts Institute of Technology
Originally presented on: 11/6/2013 15:40:00 - 16:05:00

In the pharmaceutical industry, good control of crystal size distribution (CSD) can improve process efficiency and formulation of drug product. An effective way to improve the control of CSD is through combining continuous seeding and growth with concentration control (Woo et al., 2011). To continuously produce small mono-disperse crystals, one well-studied approach is to combine solution and anti-solvent streams in a dual-impinging-jet (DIJ) mixer (e.g., Johnson and Prud'homme, 2003; Midler et al., 1994; Woo et al., 2011). The underlying principle is that, at appropriate flow rates, a DIJ mixer can generate high-intensity micro-mixing of fluids to quickly achieve a nearly homogeneous composition of high supersaturation before the onset of primary nucleation. Use of a DIJ mixer to generate crystals can sometimes remove the requirement of a milling step, which simplifies the overall manufacturing process and avoids the potential of generating an undesirable polymorphic transformation (Johnson and Prud'homme, 2003; Midler et al., 1994).

In past work, we experimentally demonstrated the use of a cooling DIJ mixer to combine hot and cold saturated solutions, rather than combining a solution with an anti-solvent, to generate seed crystals with a narrow size distribution for L-asparagine monohydrate (LAM) in aqueous solution (Jiang et al., 2012). The crystals were less than 10 microns in length, placing them in the correct size range for direct application in inhalers (Capstick and Clifton, 2012). Theoretically, if the mixing were perfect (that is, concentration and temperature were completely mixed at the molecular scale in the DIJ mixer), inspection of the phase diagram for the LAM-water system indicated that the supersaturation would not be high enough to nucleate crystals within a cooling DIJ mixer. The observed nucleation of crystals can be explained by the observation that the energy transfer rate at the interface between the impinging fluids is much faster than the mass transfer rate, so that the temperature of the hot solution near the interface can drop to the average temperature of the two solutions before its solution concentration can significantly change, resulting in the supersaturation in the hot fluid becoming sufficiently high to nucleate crystals.

While the ability to generate highly uniform-sized crystals of less than 10 microns using a cooling DIJ mixer would be a perfect technology for inhaler applications, this approach is not expected to be applicable to all combinations of pharmaceuticals and solvents. Currently, scientists/engineers have to assess whether any particular compound/solvent combination is suitable for use in a DIJ mixer by trial-and-error experimentation aided with experience (Tung et al., 2009). It would be useful at the early stage of research and development to be able to quickly identify compound-solvent combinations that cannot nucleate crystals within DIJ mixers, based on their physicochemical properties. This work answers this question for cooling jet DIJ mixers. We combine mathematical modeling of crystallization kinetics, thermodynamics, and transport with theoretical analysis to derive conditions in which a pharmaceutical/solvent combination is suitable for the nucleation of crystals within a cooling DIJ mixer. The mathematical models are in the form of partial differential equations that are simplified by exploiting symmetries and employing scaling analysis, to enable the derivation of analytical expressions that compare favorably with numerical simulations. The mathematical model and methodology are validated for the crystallization of LAM in aqueous solution.

References

Capstick, T. G. D., Clifton, I. J., 2012. Inhaler technique and training in people with chronic obstructive pulmonary disease and asthma, Expert Review of Respiratory Medicine 6(1), 91-103.

Jiang, M., Wong, M. H., Zhu, Z., Zhang, J., Zhou, L., Wang, K., Ford Versypt, A. N., Si, T., Hasenberg, L. M., Li, Y., Braatz, R. D., 2012. Towards achieving a flattop crystal size distribution by continuous seeding and controlled growth, Chemical Engineering Science 77, 2-9.

Johnson, B. K., Prud'homme, R. K., 2003. Chemical processing and micromixing in confined impinging jets, AIChE Journal 49(9), 2264-2282.

Midler, M., Jr., Paul, E. L., Whittington, E. F., Futran, M., Liu, P. D., Hsu, J., Pan, S. H., 1994. Crystallization method to improve crystal structure and size, U.S. Patent #5314506A, 11 pages.

Tung, H., Paul, E. L., Midler, M., McCauley, J. A., 2009. Crystallization of Pharmaceuticals: An Industrial Perspective. Hoboken, NJ: Wiley.

Woo, X. Y., Tan, R. B. H., Braatz, R. D., 2011. Precise tailoring of the crystal size distribution by controlled growth and continuous seeding from impinging jet crystallizers, CrystEngComm 13, 2006-2014.

(536c) Tailoring Desired Powder Properties in Pharmaceutical Development By Crystal Engineering

Wang, C., Bristol-Myers Squibb Company
Mahoney, M., Bristol-Myers Squibb Company
Shi, Z., Bristol-Myers Squibb Company
Wang, S., Bristol-Myers Squibb Company
Chan, S., Bristol-Myers Squibb
Remy, B., Bristol-Myers Squibb Company
Sfouggatakis, C., Bristol-Myers Squibb Company
Lai, C., Bristol-Myers Squibb Company
Originally presented on: 11/6/2013 16:05:00 - 16:30:00

In order to deliver pharmaceutical materials with powder properties suitable for formulation development, specific solid characteristics (among many evaluated) are determined to be critical for the improvement of process robustness in drug product processing. A model compound with a challenging crystal habit for solid dosage formulation is evaluated by multiple crystal engineering approaches to alter its physical properties. A number of material design opportunities were identified to promote the desired powder characteristics during the manufacturing process, including synthesis, crystallization, isolation, as well as post-isolation processing steps. In this study, the primary approach to enhance powder properties is driven by varying crystal shape and/or particle morphology of the subject material. Both crystal and particle morphologies can be tailored in (a) crystallization with optimized operating parameters to guide the size and morphology evolution, in (b) crystallization by an alternative means to facilitate controlled assembly formation, and in (c) direct powder processing to impact material properties. The various crystal engineering techniques encompassed in this study allow target powder properties to be met at different development stages while accommodating frequent process changes. The design strategy and the results from the applied crystal engineering techniques will be presented.

(536d) Morphology Control Through Cycles of Particle Breakage, Dissolution and Growth

Singh, M. R., Purdue University
Tung, H. H., AbbVie Inc.
Bordawekar, S., AbbVie
Ramkrishna, D., Purdue University
Originally presented on: 11/6/2013 16:30:00 - 16:55:00

Needles and plates are commonly observed morphologies in pharmaceutical crystallization which lead to difficulties in downstream processing causing higher cake resistance in filtration, resistance to free flow, and poor compressibility. The processibility of such ill-behaved crystals can be improved by reducing their aspect ratios through cycles of sequential processes involving particle breakage, dissolution and growth. These cycles are widely used in pharmaceutical industries to improve aspect ratio of crystals. The dissolution-growth cycles can improve the aspect ratios of 2D crystals only if the relative dissolution rate of the slowest face is greater than its corresponding relative growth rate. Particle breakage has the additional advantage of providing flexibility in attaining target morphologies within the feasible operating regimes. The particle breakage using either a Rotor-Stator or a Sonicator usually affects the longer dimension of the crystal, resulting in the production of fines and reduction in the aspect ratio. Successively, the dissolution of crystals will redistribute fines into the solution and may further improve the aspect ratio of crystals. Finally, the crystals are allowed to grow to increase their mass without much increase in their aspect ratio. In this work, we develop optimal cycles of particle breakage, dissolution and growth for shape control of Cholesterol crystals in a batch slurry which has the potential for considerable improvement in productivity.

(536e) Controls of Transformation Between Amorphous and Crystal Ceftiofur Free Acid

Gao, Y., National Engineering Research Center for Industrial Crystallization Technology, School of Chemical Engineering and Technology, Tianjin University
Wang, J., National Engineering Research Center for Industrial Crystallization Technology, School of Chemical Engineering and Technology, Tianjin University, Tianjin,China
Wang, Y., National Engineering Research Center for Industrial Crystallization Technology, School of Chemical Engineering and Technology, Tianjin University
Yin, Q., National Engineering Research Center for Industrial Crystallization Technology, School of Chemical Engineering and Technology, Tianjin University
Hao, H., National Engineering Research Center for Industrial Crystallization Technology, School of Chemical Engineering and Technology, Tianjin University
Originally presented on: 11/6/2013 16:55:00 - 17:20:00

Ceftiofur is a third generation cephalosporin antibiotic, which has a wide spectrum of activity against Gram-positive and Gram-negative bacteria including β-lactamase-producing strains. It can be used as veternary medicine as the sodium salt, the hydrochloride, and the free acid. Among them, the free acid was found to be the best for clinical therapy. Ceftiofur free acid could exist in either amorphous or crystalline state. Generally, amorphous ceftiofur free acid will be much easier to obtain in the separation or purification process of it .In this research, the crystallization process of ceftiofur was investigated in detail by using several PAT tools, such as ATR-FTIR, FBRM, PVM and Raman spectroscopy. The crystal morphology and the structure of the final product was also analyzed by traditional off line tools such as XRD, DSC and SNMR. The effect of operating parameter, such as solvent, temperature, dosing rate and agitation speed, on the solid state of the final product was evaluated. Optimum conditions for manufacturing both forms were successfully developed. It was also found from experiments that the amorphous state ceftiofur can transform into crystalline state in appropriate conditions although it may take a very long period. The mechanism of this transformation was investigated using the in situ tools mentioned above.

(536f) NIR-Based Process Understanding of Anti-Solvent Cocrystallization Between Carbamazepine and Saccharin

Choi, G. J.,

Originally presented on: 11/6/2013 17:20:00 - 17:45:00

Over the last decade, the cocrystal approach has been studied extensively as one of the most promising methods to improve the dissolution properties of insoluble drug substances. Cocrystal powders are typically screened or produced by mechanical grinding (wet or neat) or a solution method (evaporation or cooling). In this work, high-purity carbamazepine-saccharin (CBZ-SAC) cocrystals were produced by a novel method, anti-solvent addition. Among several solvents, methanol was found to perform well with water as the anti-solvent for the cocrystallization of CBZ and SAC. When water was added to the methanol solution of CBZ and SAC at room temperature under agitation, nucleation of CBZ-SAC cocrystals initiated within 2-3 min. Cocrystallization was complete after 30 min, giving a solid yield as high as 85% on a CBZ basis. The effects of initial concentrations, focusing on the SAC/CBZ ratio, were examined to establish optimal conditions. The whole anti-solvent co-crystallization process was monitored via in-line NIR measurements as well as at-line ATR-FTIR analysis of regularly sampled solutions. The nucleation and crystal growth of CBZ-SAC co crystals were detected by a significant increase in absorption in the range of 2400-2260 cm-1, associated with the formation of hydrogen bonds between the carbonyl group in CBZ and the N-H of SAC. When CBZ hydrates were formed as impurities during anti solvent co-crystallization, the hydrogen bonding between methanol and water was reduced greatly, primarily due to the incorporation of water molecules into the CBZ crystal lattice.

In conclusion, the whole anti-solvent cocrystallization process can be monitored for a comprehensive understanding with highly pure CBZ-SAC cocrystal powders as final products.

(548a) Improvement of Cell-Freezing Technologies and Disposable Bioreactors: Toward a Fully Closed USP Process?

Ribault, S., Merck BioDevelopment
Originally presented on: 11/6/2013 15:15:00 - 15:40:00

Biotech processes still contain a number of open and at risk transfers steps. Cell culture steps from one frozen vial to containers such as T-flasks or shake flasks is usually one of these open phases. This critical expansion step may take several days or even weeks and delay development or production timeline due to contamination. By coupling traditional technologies (bags) and new technologies (disposable bags, cases and bioreactor, Aseptic Transfer Cap), a study on bags and freezing conditions has been performed.

We have evaluated the freezing/thawing of cells in bags without DMSO removal to allow fully closed operations from thawing to the final 1250L Bioreactor inoculation. A first trial on 7 different CHO cell lines was performed to demonstrate if DMSO removal was necessary. Growth and viability were compared to the reference vial during 25 days after thawing. The best parameters have then been applied to different cell lines and media. The first part of this study demonstrated that we succeeded to freeze mammalian cells in bags with direct thawing and transfer in closed culture containers. With this technique, cell amplification timelines were reduced and the risk of cross contamination eliminated.

This first improvement was combined to single-use bioreactors that are now commonly used for process development and as seeding or production bioreactors. Disposable bioreactors have a number of advantages such as absence of cleaning, no need for sterilization, easy and fast validation, reduced labor time and increased flexibility. If the benefits associated to these equipments have been well demonstrated on more than a decade, only a few data on their scalability are published.

During the period 2010-2012, we performed a study in order to evaluate the performances of disposable bioreactors at various scale form 3L to 200L. The evaluation was performed both for seeding application and for clinical material production. A fed-batch process producing a highly glycosylated molecule was performed in different types of disposable bioreactors. The data generated allowed to conclude on the scalability and the comparability of these bioreactors. Several clinical runs at 200L and 1250L scale were performed to ensure a meaningful comparison. These process performances were also compared to glass and stainless steel bioreactors of different sizes ranging from 3.6L and 1250L. The quality of the molecule together with the molecule titer and the cell growth was compared between the different single use technologies. This study demonstrated the benefits of using disposable equipment in several key areas. The comparison of stainless steel, glass and disposable equipment showed how comparable they are.

Finally, coupling cell freezing in bag and disposable bioreactors allowed us to develop a fully closed USP process. We will extensively discuss the final set up from a technical as well as financial and organizational point of view emphasizing the various savings associated (labour, expense, training…) with single use systems and closed processes.

(548b) Effect of Anti-Apoptosis Genes On Clarification Performance

Potty, A., EMD Millipore
DiLeo, A., EMD Millipore
Xenopoulos, A., EMD Millipore
Originally presented on: 11/6/2013 15:40:00 - 16:05:00

Optimal bioreactor harvest time is typically determined based on maximizing product titer without compromising product quality. We suggest that ease of downstream purification (DSP) should also be considered during harvest. In this view, we studied the effect of anti-apoptosis genes on downstream performance. Our hypothesis was that more robust cells would exhibit less cell lysis and thus generate lower levels of cell debris and host-cell contaminants. We focused on the clarification unit operation, measuring post-clarification turbidity and host-cell protein (HCP) concentration as a function of bioreactor harvest time/cell viability. In order to mimic primary clarification using disk-stack centrifugation, a scale-down model consisting of a rotating disk (to simulate shear in the inlet feed zone of the centrifuge) and a swinging-bucket lab centrifuge was used. Our data suggest that in the absence of shear during primary clarification (typical of depth filters), a 20-50% reduction in HCP levels and 50-65% lower post-centrifugation turbidity was observed for cells with anti-apoptosis genes compared to control cells. However, on exposing the cells to shear levels typical in a disk-stack centrifuge, the reduction in HCP was 10-15% while no difference in post-centrifugation turbidity was observed. The maximum benefit of antiapoptosis genes is therefore realized using clarification options that involve low shear, < 1x10^6 W/m3 a nd minimal damage to the cells.

(548c) Process Development Comparison of Single Use Clarification Technologies for Mammalian Cell Cultures Utilizing a Conventional Two-Stage Depth Filtration, a New Optimized Depth Filter and a Novel Single-Use Fluidized Bed Centrifuge With Depth Filtration

Castano, J. C., EMD Millipore Corp.
Westoby, M. M., Biogen Idec
Elouafiq, S., Biogen Idec
McCann, D., Biogen Idec
Originally presented on: 11/6/2013 16:05:00 - 16:30:00

Single-use cell culture harvest technologies have been challenged considerably by cell culture development leading to higher productivity and cell densities with equivalent increased impurities and debris.

Conventional two-stage adsorptive depth filtration is currently used as a single use strategy for cell culture clarification, however, high density cell culture processes and increased batch volumes can impact depth filtration performance and render facility fit limitations.

A collective evaluation with multiple antibodies was accomplished to make a direct comparison of mammalian cell cultures at native and low pH (acid precipitate to induce flocculation) conditions utilizing a conventional two-stage depth filtration, two-stage single-use centrifuge with fluidized bed coupled with depth filtration and single-stage new optimized depth filtration.

This presentation will compare the results for each technology based on their performance at clarifying high-density cell cultures, reduction of impurities, recovery, and depth filtration to depict an optimal and feasible solution for single-use harvesting of high density cell cultures.

(548d) Impurity Removal in a Mab Process Using Depth Filtration Following Protein A

Schreffler, J., Morphotek, Inc.
Bailley, M., EMD Millipore
Wiltsie, E.,
Klimek, T.,
Originally presented on: 11/6/2013 16:30:00 - 16:55:00

The use of depth filters for cell culture supernatant clarification in the biopharmaceutical industry is an established and well characterized process. Recently, the usefulness of a depth filter to remove impurities when placed downstream of the capture step has been demonstrated after a change of polishing steps in one of our monoclonal antibodies led to increased levels CHO HCP, in the final product. Experiments were performed to characterize HCP levels after the addition of an EMD Millipore X0HC depth filter to this purification process post capture. The effectiveness of the X0HC depth filter to reduce HCP to comparable or better levels than those seen in the previous process was demonstrated and characterized at bench scale. This process was then successfully transferred to both the 130-L and 1,000-L scales. These initial data led to additional studies that examined whether the anion exchange flowthrough chromatography step used in this updated process was redundant. Subsequent tests showed no detectable differences for HCP, DNA, or residual protein A in the final product produced from updated purification streams with and without the anion exchange chromatography step at bench scale. These experiments suggested that the anion exchange chromatography step is a redundant step if a depth filter is used post capture step in a monoclonal antibody purification process. Additional testing, including the validation of viral clearance, further characterization of depth filtration process parameters, and scale up are necessary to validate the feasibility of this type of implementation.

(548e) Integrated Reaction Modeling and Process Evaluation in the Development of Bioconjugation Technology Platform

Ho, S. V., Pfizer Global R&D
Buckley, J. J., Pfizer
Hoeltzli, S., Pfizer Global R&D
Gamlath, C.,
Pozzo, M.,
Originally presented on: 11/6/2013 16:55:00 - 17:20:00

Bionconjugation is utilized in the formation of diverse classes of biotherapeutics such as conjugate vaccines, polymer-based biologic conjugates and antibody-drug conjugates. The development of an industrial bioconjugation process is thus inherently complex and time consuming. Important areas considered include the conjugation system (selection of carrier and active agent or payload), the type of linker used to connect the carrier to the active agent, and the conjugation reactions. Reaction kinetic modeling and reaction engineering have been found to play a beneficial role in the development of the conjugation step when used in concert with the experimental process evaluation. In this work, two similar linkers were studied, first on their hydrolytic stability then incorporated into the kinetic modeling of the overall conjugation reactions. Both a model protein and a virus like particle (VLP) were used for the conjugation. The developed model for VLP conjugation fitted the experimental data well and enables rapid evaluation of various operating schemes and process optimization.

(548f) Implementing a Flexible Recipe Management Strategy for Development of Biopharmaceuticals

Harkins, A., Janssen Pharmaceuticals
Fermier, A., Janssen Pharmaceuticals
Yeager, R.,
Originally presented on: 11/6/2013 17:20:00 - 17:45:00

Optimizing the process of developing novel biopharmaceutical products is an area of great promise to deliver value to patients and clinicians. A key fundamental prospect in optimizing that process is defining flexible standards through which we communicate our processes and associated data. The S88/S95 recipe data standards have proven a great mechanism to provide that framework in the chemical industry and applying them to Biopharmaceuticals is challenging. The challenge lies in the diverse number of systems and processes that must adopt the recipe structure including API production and Drug Product manufacturing, technology transfers, as well as process development. In this presentation we will review our progress on the implementation of said standards in the development of novel biopharmaceuticals, the challenges and benefits thereof.

(569a) Scale-Up of the Roller Compaction Process: A Practical and Dimensionless Approach

Rowe, J. M., Bristol-Myers Squibb
Crison, J. R., Bristol-Myers-Squibb
Carragher, T., Bristol-Myers Squibb
Vatsaraj, N., Bristol-Myers Squibb
McCann, R., Bristol-Myers Squibb
Nikfar, F., Bristol-Myers Squibb
Originally presented on: 11/6/2013 15:15:00 - 15:40:00

Scale-up of the Roller Compaction Process: A Practical and Dimensionless Approach

Roller compaction is a continuous dry granulation process commonly used in the pharmaceutical industry to improve powder flow and blend uniformity via powder densification. In the roller compaction process, powder is fed between two counter-rotating rolls. The friction between the powder and the rotating roll surface draws the powder through the gap between the rolls and maximum compaction (densification) occurs when the separation between the rolls is at its minimum. The resultant compacts, or briquettes, are then milled to achieve the desired granule size.

While the roller compaction process appears simple, efforts to quantitatively model the process have proven challenging due to complex material behavior in the feeding and compaction zones. To date, implementation of roller compaction models to experimental work has been limited because these models typically require large experimental data sets or obscure input parameters that are difficult to obtain experimentally. In this work, an alternative approach has been established, expanding upon the widely-used roller compaction model, Johanson's model, to enable its incorporation into a daily work-flow. The proposed method requires minimal amounts of material (< 1 kg drug blend) and only standard, routinely-measured parameters as inputs. An excellent correlation between simulated and experimental results has been achieved for placebo and active blends up to 22%w/w drug-load. Furthermore, a dimensionless relationship between key process parameters and final compact properties was elucidated. This dimensionless parameter, referred to as the modified-Bingham number (B_m*), highlights the importance of balancing yield and viscous stresses during roller compaction to achieve optimal output properties. By maintaining a constant ratio of yield-to-viscous stresses, as indicated by a constant B_m*, consistent products were attained between two scales of operation. B_m* was shown to provide guidance towards determining the design space for formulation development, as well as to facilitate scale-up development.

(569b) Extrusion Simulation and Its Application to Process Scale-Up

Zhu, Y., AbbVie
Chen, B., AbbVie
Anderson, B. C., AbbVie
Originally presented on: 11/6/2013 15:40:00 - 16:05:00

The suitability of a one-dimensional computational extrusion simulation to acquire fundamental knowledge of the extrusion processes, continuously improve commercial extrusion processes, and develop transferability between production scales has been demonstrated. By using proper inputs, including material properties and process information, a mathematical model can be tuned to match simulation results with experimental data. The correlation of the amount of a key impurity for a product vs. simulation outputs was established based on historical data. This correlation was then used to study the impact of material properties and process conditions on this quality attribute, suggesting that (1) the simulation and degradation model fit experimental data both qualitatively and quantitatively for changes in process conditions; (2) the simulation and degradation model fit experimental data both qualitatively and quantitatively for changes in material properties; (3) process conditions can be selected on the pilot scale extruder to emulate behavior on the commercial scale extruder; (4) These predictions can help to streamline experiments on both scales. The predictive model was successfully challenged at multiple degradant levels when different materials or new process conditions were used. The simulation could also have application in multiple areas including but not limited to: (1) mapping process ranges by "Virtual Design Space" exercises, (2) screening of new materials for impact on degradation, (3) linking variability from raw materials to product quality, (4) scale-up and process optimization for new products, and (5) changes in screw design to optimize extrusion equipment.

(569c) Scaling Particle Dynamics in Rotating Drum Coaters

Kumar, R., Purdue University
Wassgren, C. R., Purdue University
Ban, J., Purdue
Originally presented on: 11/6/2013 16:05:00 - 16:30:00

Scaling Particle Dynamics in Rotating Drum Coaters

J. Ban, R. Kumar, and C. Wassgren

School of Mechanical Engineering

Purdue University

Abstract

Scaling Â particle coating operations is an important issue for a number of industries, including those that process pharmaceuticals, consumer products, and chemicals. Scaling of particulate processes, and coating in particular, has been the subject of numerous experimental, computational, and analytical studies. Unfortunately, a dependable scaling method that fully translates data gathered at the lab scale to the industrial scale does not exist, especially since geometric similarity is often neglected. This study investigates how imperfect particle size scaling affects the particle dynamics and inter-particle coating variability in geometrically similar pan coaters. DEM computer simulations were used to model particle movement in five different-sized rotating cylindrical drums, resulting in five different drum to particle diameter ratios, with identical fill fractions and Froude numbers. Measurements of particle surface velocities and time spent in the coating zone were made.Â The spray coating zone residence time coefficients of variance (CoVs) were used to determine the degree of inter-particle coating variability. Surface velocity profiles show good agreement with those found in the literature; however, the velocity scaling behavior is different. The size of the coating zone was found to have no influence on the long term CoV behavior, a result that has not been described previously in the literature. Decreasing the drum to particle diameter ratio results in reduced particle mixing and smaller CoV values after the same number of drum rotations. This last finding demonstrates that the current approach to scaling without maintaining geometric similarity between the particles and the drum can result in significant errors.

(569d) Process Monitoring of a Full Continuous Pharmaceutical Tabletting Line: Mass & Energy Balance Analysis

Mortier, S. T. F. C., Ghent University
Nopens, I., Ghent University
De Beer, T., Ghent University
Gernaey, K. V., Technical University of Denmark (DTU)
Originally presented on: 11/6/2013 16:30:00 - 16:55:00

Nowadays, pharmaceutical production processes are undergoing major changes, and there is a clear trend towards increased use of continuous production processes [1]. The focus of our ongoing work is on the formulation of the Active Pharmaceutical Ingredients (APIs) into final drug products. The evolution from batch to continuous production is also of increasing interest in formulation operations [2-3]. Whereas in traditional batch processes product quality evaluation mostly relies on off-line, time-consuming and, hence, less efficient laboratory testing [4], the continuous approach possesses clear advantages as the efficiency can be improved by relying on in-line measurements and real-time adjustment of sensitive process variables. Monitoring and controlling the process during operation can be much more economical through significant decreases in product loss [4-5]. By using the on-line monitoring to support the implementation of real-time release in the production of pharmaceutics, pharmaceutical companies does not have to wait anymore till the product is approved by the quality lab.

The ConsiGma™ (Collette™, GEA Pharma Systems, Wommelgem, Belgium) is a continuous from-powder-to-tablet manufacturing line intended for tablet production. The continuous line consists of three parts: a continuous twin screw granulator (high shear), followed by a six-segmented fluidized bed dryer system and a discharge system prior to the tabletting press. During continuous operation a large number of process variables are continuously logged. These variables can be used by the operator to verify smooth process behavior. However, they can be analysed in more detail as well, in order to improve process operation. In this contribution, a mass and energy balance verification over the fluidized bed dryer unit is performed, which provides further insight in the process.

The mass balance can be used to calculate the moisture content at the end of the drying process. As such, by making a mass balance a time-consuming off-line Karl Fisher titration could be avoided or done on a less regular basis. The mass balance is based on the changing properties of the drying air passing the dryer unit from inlet to outlet and accounts for the evaporation process. The energy balance is interesting for two reasons; it can be used to compare the measured gas outlet temperature with the calculated value based on logged variables. A good match between prediction and measurement will increase trust in the correctness of the mass balance, and demonstrates process understanding. On the other hand the energy balance can be helpful to predict the gas temperature in different horizontal sections of the drying system. This information is useful to combine in a later stage with information about the trajectory of granules through the system.

The fluidization behavior is investigated in another part of the project using Computational Fluid Dynamics (CFD). The CFD-model can give information about the bed height of the fluidized bed; evaporation of water only takes place in this part of the dryer, which is the most determining factor for the gas temperature. The temperature profile will be based on several model assumptions, which is acceptable since experimental measurements are time-consuming and laborious.

Experimental data revealed that a calibration was needed in order to calculate the evaporation rate as a consistent offset between prediction and measurement occurred. A linear regression was implemented in the model to quantify this offset. A linear error propagation of measurement errors was pursued to account for their effect on output uncertainty.

Based on several datasets, it can be concluded that it is indeed possible to use the standard logged measurements to monitor the process. As such, an important conclusion is that expensive probes (e.g. NIR [6]) are not always needed to implement in the final production system. However, they are still extremely useful in process development to support development of detailed process knowledge.

[1] K. Morris et al., Advances in pharmaceutical materials and processing, Pharmaceutical Science 1 (1998) 235-245.

[2] C. Vervaet, J.P. Remon, Continuous granulation in the pharmaceutical industry, Chem. Eng. Sci. 60 (2005) 3949-3957.

[3] H. Leuenberger, New trends in the production of pharmaceutical granules: the classical batch concept and the problem of scale-up., Eur. J. Pharm. Biopharm. 52 (2001) 279-288.

[4] S.T.F.C. Mortier et al., Mechanistic modelling of fluidized bed drying processes of wet porous granules: A review, Eur. J. Pharm. Biopharm. 79 (2011) 205-225.

[5] H. Leuenberger, New trends in the production of pharmaceutical granules: batch versus continuous processing., Eur. J. Pharm. Biopharm. 52 (2001) 289-296.

[6] P. Frake et al., Process control and end-point determination of a fluid bed granulation by application of near infra-red spectroscopy, Int. J. Pharm. 151 (1997) 75-80.

(569e) Use of Design of Experiments (DoE) and Empirical Modeling in Liquid-Liquid Extraction Design Space Development

Ferreira, G., Bristol-Myers Squibb
Hallow, D., Bristol-Myers Squibb Company
Fenster, M., Bristol-Myers Squibb Company
Mack, B. C., Bristol-Myers Squibb Company
Lou, S., Bristol-Myers Squibb Company
Domagalski, N., Bristol-Myers Squibb Company
Hobson, L., Bristol-Myers Squibb Company
Originally presented on: 11/6/2013 16:55:00 - 17:20:00

A robust quality control strategy for the production of active pharmaceutical ingredients (APIs) and related intermediates is an essential element of pharmaceutical process development. Although numerous examples for reactions and crystallizations exist in the literature, techniques employed during ubiquitous operations such as liquid - liquid extraction are limited. This presentation will address that gap by reviewing the creation of a control strategy to purge a key impurity from an intermediate via liquid - liquid extractions. An important element of laboratory studies proved to be application of multiple rounds of fractional factorial design of experiments (D o E). The collected data was used to both construct and verify an empirical model and later construct a response surface. A five factor, multivariate design space was identified and formed the basis for the quality control strategy. The strategy has been successfully demonstrated in numerous pilot plant scale campaigns and validated at commercial manufacturing.

(569f) Modelling of Fluid Bed Drying At Different Production Scales in Pharmaceutical Drug Manufacture

He, X., Boehringer Ingelheim
Pinto, M., Process Systems Enterprise
Originally presented on: 11/6/2013 17:20:00 - 17:45:00

Fluid bed drying is commonly used to dry material subsequent to wet granulation in pharmaceutical drug product manufacture. The key decision variables in the fluid bed drying process are the flowrate and temperature of the drying air. A constraint is usually imposed limiting the temperature of the solid material in order to prevent undesirable physical and chemical changes. In operating batches, the goal is to achieve a desired solid moisture content as quickly as possible.

In this presentation, a model is developed of a fluid bed drying process based on one published in the literature (Burgschweiger et al, 1999; Burgschweiger and Tsotsas, 2002). The model is validated against data obtained using placebo material at the lab scale. The model is able to predict batch drying profiles (solid temperatures and moisture contents) for different initial moisture contents and particle sizes. It is observed that depending on the nature of the material being dried, externally limited drying (the drying rate is limited by mass transfer to the vapour phase), internally limited drying (the drying rate is limited by mass transfer from inside the particle to the surface) or both may occur.

After validating the model using data on the lab scale, the model is applied to data obtained from production of real drug material at the production scale. It is observed that the model is able to predict batch drying profiles at the larger scale as well. This mechanism based model could be used to gain process understanding and control in scaling up the fluid bed drying process.

References

- Burgschweiger J, Groenewold H, Hirschmann C, Tsotsas E (1999) From hygroscopic single particle to batch fluidized bed drying kinetics. The Canadian Journal of Chemical Engineering 77, 333-341
- Burgschweiger J, Tsotsas E (2002) Experimental investigation and modelling of continuous fluidized bed drying under steady-state and dynamic conditions. Chemical Engineering Science 57, 5021 – 5038

(586b) Bioproduction and Formulation Patents - Opportunities and Challenges for Biotherapeutic Development

Calvo, P., Sterne, Kessler, Goldstein & Fox

Patents that cover bioproduction methods and formulations are important for patent life cycle management of all biotherapeutics. However, there are patentability challenges associated with obtaining patents that cover aspects in addition to the drug composition. Thus, the importance of patents, as well as trade secrets, and an understanding of the opportunities/challenges they present is invaluable. Moreover, recent changes to the US patent laws have introduced new avenues to challenge patents that were not previously available. These new inter partes proceedings may provide earlier resolution for patent conflicts, as well as bring about settlements that could avoid more costly litigation altogether. A clear understanding of patent issues is absolutely vital in all areas of therapeutic development, especially in the area of biosimilars. This presentation will use existing patents as examples to outline the opportunities and challenges for both originator and follow-on companies as they create portfolios to protect their compound, and/or plan design-around strategies to challenge these so-called second generation patents.

(586c) Selection of Polymorphic Forms Via Continuous Kinetically Controlled Msmpr Crystallization

Ferguson, S. T., Massachusetts Institute of Technology
Lai, T. T. C., Massachusetts Institute of Technology
Palmer, L.,
Trout, B. L., Massachusetts Institute of Technology
Myerson, A. S., Massachusetts Institute of Technology

This work presents the first study of the mechanism controlling crystallization of polymorphs in continuous MSMPR type crystallizers. Crystallization is the dominant purification and separation technique used in the pharmaceutical industry, with more than 90% of small molecule pharmaceutical products utilizing a crystallization step in their production [1]. For this reason the understanding of polymorphism in continuous crystallization is vital for the future implementation of continuous processing in pharmaceutical manufacturing. It was found that if the MSMPR crystallization used in this study was operated at a steady state so that supersaturation exists with respect to both the metastable (alpha) and stable (beta) polymorphs for L-glutamic acid (a model monotropic organic system). The controlling mechanism as to which polymorph was produced was the relative rates of secondary nucleation of the two polymorphs.

This is due to the fact that when solution mediated polymorphic transformation is eliminated, the ability to sustain the number of particles in the crystallizer at a given residence time becomes the controlling factor as to which polymorph or the relative amount of each that are produced. This enabled selective crystallization of either the meta-stable or stable form by a purely kinetically controlled crystallization, and may provide the first robust and scalable method for isolation of metastable forms on an industrial scale. The use of mother liquor recycle enabled the yield in either of these scenarios to at least match that of the equivalent batch crystallization [2].

[1] Variankaval, N., Cote, A. S., Doherty, M. F. (2008). From Form to Function: Crystallization of Active Pharmaceutical Ingredients. *AIChE Journal* 54(7): 1682-1688.

[2] Wong, S.Y., Tatusko, A.P., Trout, B.L., Myerson, A.S. (2012). Development of Continuous Crystallization Processes. Using a Single-Stage, Mixed-Suspension, Mixed-Product Removal Crystallizer with Recycle. Cryst. Growth & Des. 12(11), 5701-5707.

(586v) An Effective and Facility Compatible Cleaning in Place (CIP) Procedure for Affinity Resin in Monoclonal Antibody (Mab) Purification

Wang, L., Bristol-Myers Squibb
Dembecki, J., Bristol-Myers Squibb
Jaffe, N., Bristol-Myers Squibb
Cai, H., Bristol-Myers Squibb
Sparks, C., Bristo-Myers Squibb
Zhang, J., BMS
Laino, S., BMS
Russell, R. Jr., Bristo-Myers Squibb
Wang, M., Bristol-Myers Squibb

Cleaning-in-place (CIP) for column chromatography plays an important role in therapeutic protein production. A robust and efficient CIP procedure improves column life time and increase cost effectiveness of the purification processes, particularly for those using highly expensive affinity resins. Cleaning efficiency, resin compatibility, and facility compatibility are the three major aspects to consider in CIP process design. Cleaning affinity resin with sodium hydroxide along with high concentration of salt is one of the most popular cleaning procedures used in biopharmaceutical industries. However, high concentration sodium chloride could cause corrosion in the stainless steel containers used in large scale manufacture. Therefore, it is challenging to apply this cleaning procedure into commercial scale purification due to facility compatibility. Here we introduce an effective and environmental-friendly cleaning procedure which is suitable for large scale affinity chromatography. A high throughput screening method was developed in 96-well plate format for selection of the most effective cleaning reagent from ten candidate solutions which meet facility compatibility. Resin compatibility of selected cleaning reagents was also evaluated. The optimized cleaning strategy was comprehensively investigated in resin life cycle study. The new cleaning regime is able to prolong resin life time and consistently deliver high purity drug products.

(586d) Operational CO_2 Recycling Study On a Preparative Supercritical Fluid Chromatography System

Whelan, J. II,
Wang, Z.,
Baugher, J., Waters Corporation

Supercritical Fluid Chromatography(SFC) is a useful tool for analytical and large scale separations. SFC is most often used with carbon dioxide as a mobile phase and an organic modifier such as some type of organic alcohol. It has some significant advantages over standard Liquid Chromatography methods such as less pressure drop across the columns, faster column equilibration, faster method development, higher efficiency separations and significantly less generation of hazardous waste. Some of the main advantages for preparatory Supercritical chromatography include solvent waste reduction, facilitated product recovery, lower solvent cost and the possibility for recycling.

One of the advantages of using Supercritical Fluid Chromatography for large scale separations is the ability to recover the CO_2 and recycle it concurrently with executing the process. In the past some recycle presentations have been done, but typically with only one peak being used as an example. Contained in this study we have demonstrated CO_2 recycling with multi-compound mixtures that yield two peaks or more peaks. In using multi-compound mixtures a more accurate picture of the capability of the Recycler being used is portrayed. The body of work will include Retention Time RSD% and effects of slight co-solvent/modifier changes with regards to retention time. The retention times will be displayed over varying conditions and also two unattended overnight trials. Recycle efficiency for the methods and conditions used in the multi-component tests will also be discussed. Our data has shown high recycling efficiency(RE) with the designed system and excellent chromatographic performance for the separations.

(586w) Assessing the Relative Cleanabilities of Bioprocess Soils

Wood, J.,
Arroyo, A., Genentech, Inc.

Bench-scale methods for assessing the relative cleanabilities of process soils are useful tools for applying risk-based approached to full-scale cleaning validations, troubleshooting cleaning issues and developing cleaning strategies. In this study, a spreading method for controlling soil size and shape during sample preparation and an image-analysis technique for constructing "cleaning profiles" from which relative cleanability may be determined were explored. Variables useful for studying spreading and cleanability were defined – Spreading Ratio (SR) and Specific SR (SSR), which quantify surface-area coverage and average area density, respectively. Using these methods, strategies for comparing soil spreading on various materials of construction and determining relative cleanabilities among larger groups of soils are presented.

(586a) Development and Evaluation of Nanostructured Formulations Containing 8-Methoxalen

Barradas, T., Macromolecules Institute - Federal University of Rio de Janeiro
Silva, K. G., Pharmacy School - Federal University of Rio de Janeiro
Pinto, A. P. D. S.,
Mansur, C. E., Federal University of Rio de Janeiro

Vitiligo is a disorder characterized by cutaneous depigmentation of the epidermis and predominantly affects the face area, causing serious emotional complications in patients. The main topical treatment for vitiligo is based on a therapy with a psolaren such as 8-methoxalen (8-MOP) associated with ultraviolet a radiation (UVA). However, this approach has presented some disadvantages, such as frequent local phototoxicity and slow response. Nanostructured systems are able to promote controlled release and favor the folicular penetration, which cannot be achieved with conventional dosage forms. Micelles can entrap drugs and thereby control the release of such molecules and promote skin retention of drugs. Considering the advantages of nanoemulsified systems for the administration of poorly soluble drugs, to investigate the feasibility of introducing the 8-MOP in nanoemulsions may be an interesting approach for developing a topical formulation.

In this work we will discuss the development and characterization of nanostructured formulations for prolonged release of (8-MOP). The first step consisted in the selection of the oily phase for the preparation of nanoemulsion. The oil has been chosen after a study of the solubility of 8-MOP in several different oils. Then, polymeric surfactants have been studied at different concentrations, for the preparation of nanoemulsions containing the oily phases preselected. The nanoemulsions were prepared by the energy input of mechanical devices such as ultrassonicator. All systems were characterized regarding the average droplet diameter and storage stability by dynamic light scattering using equipment Zetasizer Nano ZS (Malvern Instruments Ltd., UK). The morphology of all formulations was accessed by Transmission Electron Microscopy. The *in vitro*drug release studies have been carried on a vertical diffusion system composed of a donor compartment and receptor compartment filled with a receptor solution. Between both compartments, a semi-synthetic membrane composed by cellulose acetate was placed. Samples were collected at regular time intervals and analyzed by HPLC to assess the amount of 8-MOP released. The results indicate that the formulations developed in this study represent a promising approach for the topical treatment of located vitiligo.

(586ab) An In-Silico Approach for Investigating Molecular Interactions in Carbohydrate-Water Mixtures

Gupta, J., University of the Sciences
Nunes, C., Bristol Myers Squibb
Jonnalagadda, S., University of the Sciences

PURPOSE

The objectives were (i) to develop a molecular dynamics (MD) based approach for determination of radial distribution function (RDF) and (ii) to delineate the specific interactions in a model carbohydrate (sucrose, SUC) and water system.

METHODS

MD construct comprised of periodic amorphous cells of SUC containing 0-5%w/w water. The COMPASS force-field was utilized. The cells were subjected to energy minimization (conjugate gradient), and then equilibration (2ns) using canonical (NVT) and isothermal-isobaric (NPT) ensembles at 440K. Each system was cooled at 5K intervals spanning temperature range of 440-265K. The configurations corresponding to the average density values, obtained from NPT, were subjected to NVT simulations (200ps). The RDF was calculated from the terminal 50ps simulation time for water-oxygens (O_w) around SUC hydroxyl-oxygens (O_{HS}) and around SUC acetal-oxygens (O_{AS}).

RESULTS

The RDF is measure of probability of finding a given pair of atoms at a distance *r* with respect to bulk phase in a random distribution. For O_{HS} -O_w pairs, a sharp primary and a broad secondary peak were observed at 2.8°A and 5.5°A, respectively. The sharp peak suggested strong hydrogen-bond interactions and was ascribed to the first hydration shell with water molecules structurally arranged within the shell (width 1Â°A).Â The broad peak suggested the presence of second hydration shell wherein the water molecules were less structurally arranged.

A lesser degree of hydrogen-bonding was observed for O_{AS} -O_w pairs as evident by the presence of distorted/less pronounced first peak. This was attributed to O_{AS} constrained ring geometry, bulky neighboring groups and less accessibility to surrounding water molecules. The RDF data suggests that water disrupts H-bond network between SUC-SUC and forms hydrogen-bond between SUC-water thereby destabilizing SUC. The magnitude of interactions was dependant temperature of the system and correlated to molecular mobility.

CONCLUSIONS

The RDF measurements revealed strong hydrogen-bond interactions between SUC hydroxyl-oxygens and water-oxygen. Steric effects led to weak interactions between SUC acetal-oxygens and water-oxygen. Molecular interactions play critical role in stabilization or phase separation of multi-component amorphous systems. MD is thus a powerful tool for assessing the stability risks during development of an amorphous drug product.

(586ac) Effect of Plasticizer On Glass Transition Temperature of Amorphous Pharmaceuticals - A Computational Study

Gupta, J., University of the Sciences

Nunes, C., Bristol Myers Squibb

Jonnalagadda, S., University of the Sciences

PURPOSE

The objectives were: (i) to develop a molecular dynamics (MD) based technique for predicting glass transition temperature (T_g) and (ii) to computationally evaluate the effect of plasticizer (water) on T_g of a model amorphous compound (sucrose, SUC).

METHODS

The periodic *in-silico* models of amorphous SUC containing 0%, 3% and 5%w/w water were constructed at 298K. COMPASS forcefield was utilized for the MD simulations. The cells were energy minimized (steepest descent, conjugate gradient), and then equilibrated (2ns) using canonical (NVT) and isothermal-isobaric (NPT) ensembles at 440K. Each system was quench cooled from 440K to 265K at intervals of 5K. The specific volumes (V) were computed using the cell densities and plotted as a function of the corresponding temperature (T). Three independent simulations for each system were performed to improve sampling statistics.

RESULTS

The computed specific density of the amorphous SUC cells at 298K was in agreement with experimentally determined value. A gradual increase in density of the amorphous cells, along with corresponding decrease in specific volume, was noted as each system was quenched. As the system approached glassy state, a discontinuity in V-T curve was observed. This characteristic ?kink' represents the transition from less stable ?rubbery state' to a kinetically stable ?glassy state'. Statistical regression analysis of the data corresponding to the glassy and the rubbery region identified the T_g to be 367K. The RSD for data points corresponding to the rubbery phase was greater than that for the glassy state, attributable to greater molecular mobility in the rubbery state. Increasing the water content progressively decreased the system T_g. The MD computed T_g values for amorphous SUC containing 3% and 5% w/w water, were 352K and 343K respectively.

CONCLUSIONS

The T_g of amorphous SUC was successfully computed by means of full atomistic based MD simulations. Enhanced mobility in the rubbery state was also manifested as increased scatter of data points in the high temperature regimes. The simulated plasticization effect corroborated well with theoretical model predictions and reported experimental data. The technique can thus provide a molecular level insight in to the physical stability of amorphous formulations during product development.

(586r) Simulation and Global Sensitivity Analysis of Pharmaceutical Processes for Solid Drug Manufacture

Rogers, A., Bristol-Myers Squibb Co.
Ierapetritou, M. G., Rutgers, The State University of New Jersey
Inamdar, C., Rutgers University

There are significant economic incentives to transition from batch to continuous processing of pharmaceutical products (1). However there are several challenges to be overcome in order to implement continuous manufacturing systems for solids-based processes. Among these is the need to develop increased understanding of continuous processes and their sensitivity to variability in input materials, processing equipment and operating conditions. The creation of predictive process models can play an important role in developing the requisite process understanding. Process models can be used to reduce experimental requirements, enhance process understanding, and develop control strategies. Modeling is also an important component of the Quality by Design (QbD) paradigm described by the ICH Q8 guidance for pharmaceutical development. (2)

In recent work a computationally efficient approach for the simulation of integrated pharmaceutical processes based on the use of an equation-oriented process simulator (gProms®) has been presented. Within this framework both mechanistic and semi-empirical models for particulate processes havebeen implemented. (3) In the current work this framework is used to model a process for the continuous production of pharmaceutical tablets via direct compaction. Direct compaction is the most straightforward approach to tablet production and is thus an appealing manufacturing route if it can be successfully implemented for the product of interest. Because the direct compaction process does not include a granulation step, the physical properties of the raw materials have the potential to significantly impact performance during compaction. Content uniformity may also be quite sensitive to powder properties and the performance of unit operations such as blending. Therefore it is important to understand the impact of material properties and unit operation design and operating parameters on process performance if direct compaction is to be successfully implemented. To this end sensitivity analysis can be conducted with the dual objectives of model improvement and process understanding.

Sensitivity analysis is an important component of modeling that generally occurs in tandem with process development. Sensitivity analysis can be used to identify critical process parameters and to examine the validity of assumptions underlying the developed process models. In the current work, global sensitivity analysis is used to identify gaps in the existing process model with the goal of defining specific areas where model improvement can be effected. In addition, global sensitivity analysis of the integrated flowsheet simulation is used to assess the impact of variability in input material properties and process design and operating parameters on product quality.

In the current work a global sensitivity analysis is conducted which considers the effect of 22 uncertain design, operating and raw material properties parameters on 19 responses including indicators of unit operation performance and of product quality. A variety of global sensitivity analysis techniques are implemented, including regression based and variance based methods. The partial rank correlation coefficient, a regression based sensitivity metric calculated from rank transformed data, has been used for preliminary sensitivity analysis. PRCCs are most accurate as an indicator of sensitivity when the inputs are not highly correlated and the responses vary monotonically with the inputs. (4) In this case the majority of input-response relationships are monotonic or nearly monotonic, but some of the model inputs are correlated. Thus PRCCs alone may not be sufficient to guarantee accurate assessment of sensitivities. In addition, the partial rank correlation coefficient does not provide an indication of the extent to which interactions among parameters may influence the process. This can be addressed using the extended FAST (eFAST) method, which gives variance based sensitivity and total sensitivity indices. (5), (6) These sensitivity metrics are valid for input-output relationships that range from linear to nonlinear and non-monotonic and are thus more widely applicable than PRCCs. In addition, the total sensitivity indices indicate the sensitivity of a model output to all higher order interactions of a process parameter with all other inputs. However, eFAST does not provide information about specific interacting inputs which give rise to the total sensitivity indices. For this, random sampling HDMR (RS-HDMR) can be used. RS-HDMR yields first order sensitivity indices and second order indices for specific parameter interactions. (7), (8) (9) In this case RS-HDMR is used as a supplement to the extended FAST method in order to determine if there are any specific parameter interactions that contribute significantly to total sensitivity for a given output.

The results of global sensitivity analysis have been used to suggest potential model improvements and to identify process parameters that may be deemed critical to ensuring product quality. For instance, the lubricant concentration does not significantly affect the simulated tablet hardness. However it is known in practice that both lubricant and API concentration can result in varying tablet properties. Therefore a model modification to account for the impact of tablet composition on hardness is suggested to improve the predictive ability of the tablet compaction model. Tablet weight and composition are found to be highly sensitive to the API and excipient feeder parameters and raw material bulk densities. Thus sensitivity analysis can also contribute to determination of critical process parameters. These variables should be manipulated in order to successfully implement a control strategy for tablet API concentration.

1. *Economic Analysis of Integrated Continuous and Batch Pharmaceutical Manufacturing: A Case Study.* **Schaber, S.D., Gerogiorgis, D.I., Ramachandran, R., Evans, J.M.B., Barton, P.I., Trout, B.L.**17, 2011, Ind. Eng. Chem. Res., Vol. 50, pp. 10083–10092.

2. *ICH Harmonised Tripartite GuidelinePHARMACEUTICAL DEVELOPMENT Q8(R2) Current Step 4 version.* s.l. : International Conference on Harmonisation of Technical Requirements for Registration of pharmaceuticals for human use, August 2009.

3. *Reduced-order discrete element method modeling.* **Boukouvala, F., Gao, Y., Muzzio, F., Ierapetritou, M.G.**Chemical Engineering Science, p. Article in Press.

4. **Saltelli, A.K., Chan, K., Scott, M.** *Sensitivity Analysis.* s.l. : John Wiley & Sons Inc., 200.

5. *An alternative way to compute Fourier Amplitude Sensitivity Test (FAST).* 1998, Comput. Statist. Data Anal., Vol. 26, pp. 445-460.

6. *A quantitative, model independent method for global sensitivity analysis of model output.* **Saltelli, A., Tarantola, S., Chan, K.**1999, Technometrics, Vol. 41, pp. 39-56.

7. *Regularized random-sampling high dimensional model representation (RS-HDMR).* **Li, G., Rabitz, H.**3, 2008, Journal of Mathematical Chemistry, Vol. 43, p. 1207.

8. *Global uncertainty assessments by high dimensional model representations (HDMR).* **Li, G., Wang, S.W., Rabitz, H., Wang, S., Jaffe,P.**2002, Chemical Engineering Science, Vol. 57, pp. 4445 – 4460.

9. *GUI–HDMR – A software tool for global sensitivity analysis of complex models.* **Ziehn, T., Tomlin, A.S.**2009, Environmental Modelling & Software, Vol. 24, pp. 775-785.

10. *Computer-Aided Flowsheet Simulation of a Pharmaceutical Tablet Manufacturing Process Incorporating Wet Granulation.* **Boukouvala, F., Chaudhury, A., Sen, M., Zhou, R., Mioduszewski, L., Ierapetritou, M.G., Ramachandran, R.**2013, J Pharm Innov, Vol. 8, pp. 11-27.

11. *An integrated approach for dynamic flowsheet modeling and sensitivity analysis of a continuous tablet manufacturing process.* **Boukouvala, F., Niotis,V., Ramachandran, R. Muzzio, F.J., Ierapetritou, M.G.**2012, Computers and Chemical Engineering, Vol. 42, pp. 30-47.

12. *Sensitivity analysis for nonlinear mathematical models.* **Sobol, I.M.**1993, Mathematical Modelling & Computational Experiment, Vol. 1, pp. 407-414.

13. *A Global Sensitivity Study of Sulfur Chemistry in a Premixed Methane Flame Model Using HDMR.* **Ziehn, T., Tomlin, A.S.**s.l. : Wiley Periodicals, Inc., 2008, Int J Chem Kinet, Vol. 40, pp. 742–753.

14. *A Quantitative Model-Independent Method for Global Sensitivity Analysis of Model Output.* **Saltelli, A., Tarantola, S., Chan, K.P.-S.** 1, 1999, Technometrics, Vol. 41, pp. 39-56.

(586f) Using On-Line Mass Spectrometry to Predict the End Point During Drying of Pharmaceutical Products

Dodda, A. G., University of Massachusetts, Amherst
Saranteas, K., Sunovion Pharmaceuticals
Henson, M. A., University of Massachusetts Amherst

Downstream processing of Active Pharmaceutical Ingredients (APIs) typically involves extraction, crystallization, filtration and drying. In addition to being the most energy intensive unit operation, drying is often a major manufacturing bottleneck due to its relatively long processing times. Furthermore, drying conditions must be carefully selected and maintained to avoid problems such as API degradation. As a result, the optimization of dryer operation is needed to ensure product quality and minimize operating time. A key requirement for any optimization strategy is the ability to determine the drying end point, the time at which all solvent has been evaporated from the solid cake. While a variety of mathematical models have been developed to describe drying processes, available models are not sufficiently accurate to predict the end point and/or are too complex to be used for real-time applications in a pharmaceutical manufacturing environment. An alternative approach is to develop Process Analytical Technology (PAT) tools that take advantage of emerging on-line measurement capabilities. Near Infrared (NIR) spectroscopy and Mid Infrared (MIR) spectroscopy have been proposed as suitable methods for monitoring dryer performance and detecting the end point. However, these approaches require the development of chemometric models to map the measured spectra to solvent concentrations. Moreover, spectroscopic methods are only capable of measuring gas phase solvent concentrations, which can be quite different than solvent concentrations in the cake due to the gas residence time of the drying oven.

In this presentation, we describe the development and testing of a novel method for determining the end point for pharmaceutical dryers based on on-line mass spectroscopy. The proposed method offers several advantages over existing spectroscopic methods, including the ability to detect when the cake is dry from vapor phase measurements and very simple implementation that does not require chemometric models. The drying end point for each solvent is determined as the time at which the gas phase solvent concentration measurement from the mass spectrometer converges to a predicted value computed from a solvent mass balance on the oven assuming zero flow rate from the cake. The method was tested on a laboratory scale vacuum dryer over a range of temperatures and pressures using glass beads with three different particle sizes. Drying end points were detected for acetone, methanol and methanol-MTBE solvents well before the unprocessed gas phase solvent concentration measurements suggested that drying was complete. We found that the drying rate increased and the end point was reached more quickly as the mean bead size increased. The method was validated by performing Loss On Drying (LOD) experiments for one combination of pressure, temperature and bead size. Application of the method to an API with methanol-MTBE solvents produced a substantially reduced drying rate compared to the glass bead, most likely due to interactions between the API and solvents. We concluded that the proposed method represents a powerful Quality by Design (QbD) approach for pharmaceutical drying processes.

(586ad) Effect of Double Layer Formulation On Crystallization of Naproxen/Polyethylene Glycol Dispersion System

Hsu, H. Y., Purdue University

Solid dispersions are an increasingly important formulation approach to improve the dissolution rate and apparent solubility of poorly water soluble compounds. The matrix of hydrophilic polymer confines movement of the drug and depress its crystallization, which decreases the activation energy of dissolution. However, there are still issues of this approach. Two of the main obstacles to its application are difficulty of manufacturing and poor chemical and physical stability.

In this study, molten mixture of naproxen (NAP) and polyethylene glycol (PEG) 3350 was dispensed by a syringe pump, covered by another layer of PEG with different molecular weight (PEG2000, PEG3350, PEG6000, and PEG8000), and stored in environment with different relative humidity (RH). The extent of crystallization of NAP within each sample was assessed and compared. The practicability of combining drop printing technique and multilayer formulation to simplify manufacturing process, increase stability, and even control the releasing behavior of solid dispersion was examined.

(586ae) Disposable Bags in Biopharmaceutical: Approaches to Storing Difficult Solutions

Lee, K.,
Mahajan, E.,
Arroyo, A., Genentech, Inc.

Disposable bags have emerged as an attractive alternative to traditional stainless steel and glass containers used in biomanufacturing because they offer potential technical and economic advantages, as well as flexibility in operations. However, as these disposable bags are used in an increasing number of applications, previously identified risks are becoming a reality that must be managed and new challenges are arising. Bags that are acceptable for aqueous buffers in an upstream step are not necessarily acceptable for other process streams, especially if the step is closer to the final product. The purpose of this presentation is to discuss the storage of "difficult" solutions, like surfactants, in bioprocess bags for product contact applications during the manufacture of clinical or commercial biologics. The discussion will include a case study of storage of diluted Polysorbate buffer, highlighting the challenges, evaluation process, and selection of an appropriate bag for the application. Key elements of the project that will be discussed in detail are (1) qualification of the bag film and (2) assessment of process compatibility.

(586h) Operational Aspects Of Continuous Pharmaceutical Production

Mitic, A.,
Gernaey, K. V., Technical University of Denmark (DTU)
Johansen, K. D., Technical University of Denmark
Skovby, T., Lundbeck A/S

Background

Organic synthesis is essential for the production of an important class of pharmaceuticals. The implementation of organic chemistry on an industrial scale is a great challenge. To date, pharmaceutical manufacturing is mainly based on batch and semi-batch processes. However, besides flexibility and versatility of batch vessels there are many disadvantages. The main problems are linked to long reaction sequences, the occurrence of non-uniform conditions inside the vessels, the limited potential to apply process control and as a consequence difficulties in automating the production, and complicated cleaning procedures. Additionally, Process Analytical Technologies (PAT) cannot reach the full benefits in this type of production. Therefore, a solution to those problems might be moving to continuous production mode and establishing in-line process monitoring and control.

Introduction

In theory, continuous production conserves energy and natural resources leading to more eco-friendly and economical processes. Furthermore, due to the high degree of automation and process controllability of continuous processes, PAT may be implemented in a very efficient way. However, when switching an organic synthesis-based process from batch to continuous operation, it is important to realize that not all reactions are suited to be operated in a continuous mode. Slow reactions, for example, result in operational problems when operated in continuous mode.

Therefore, acceleration of slow chemical reactions is the main focus here in order to incorporate also this type of reactions in a continuous process efficiently. As an example process, the synthesis of zuclopenthixol – a product of H. Lundbeck A/S – is studied, and more specifically one production step – the dehydration of "Allylcarbinol" (9-Allyl-2-Chlorothioxanthen-9-Ol) to cis/trans - "Butadiene" (cis/trans - 9H-Thioxanthene, 2-chloro-9-(2-propenylidene)-(9CI)).

Despite acceleration and continuous manufacturing, increased stereo-selectivity is needed because just the cis-isomer will lead to the medically active product. Hence, the cis isomer of "Butadiene" is a precursor to cis-clopenthixol (zuclopenthixol) via the consequent hydroamination reaction with 1-(2-Hydroxyethyl)piperazine. Trans isomers are separated afterwards in a sophisticated and expensive way and will undergo the isomerization procedure.

Results and discussions

Zuclopenthixol synthesis could be divided in two parts. The first part involves fast and exothermic chemical reactions, such as Grignard reaction and hydrolysis whereas the second part includes slow and endothermic chemical reactions – dehydration and hydroamination. The main focus here is on the dehydration reaction which is traditionally performed in batch mode in the Lundbeck A/S company. Acetic acid anhydride and acetyl chloride are used as chemical catalyst, as well as toluene as a solvent.

However, long reaction times and low selectivity influenced the development of new production processes. Furthermore, the previous reaction steps (Grignard alkylation and hydrolysis) are performed in THF and thereby avoidance of the solvent swap procedure was a desired choice. Hence, the simplified synthetic route to cis/trans "Butadiens" is proposed involving usage of Brønsted acids, such as hydrochloric or sulfuric acids. The first tests were performed in batch mode leading to the total conversion of "Allylcarbinol" and producing a mixture of equal amounts of cis/trans "Butadienes". As a result, a cheaper and a very efficient synthesis step was developed, but the reaction time was still relatively high in view of performing the chemical reaction in continuous mode.

Hence, following the Arrhenius equation led to the conclusion that doubling of the reaction temperature could lead to a significant acceleration of this chemical reaction. As a consequence, a high-pressure process was developed by increasing the boiling point temperature of THF to 120°C. For this purpose, a back-pressure regulator was used. The reaction time was consequently decreased from 2h to just 3 min. Furthermore, in-/at-line process monitoring was established by using a NIR spectrophotometer, as well as off-line analysis by applying HPLC.

Despite development a fast manufacturing procedure, yield of the desired product is still low. Therefore, an increase of the fraction of the cis-isomer is necessary. For this purpose, different Lewis acids and bases were tested, as well as special dehydration agents. As result, an increased amount of the cis-isomer was achieved by using thionyl chloride and triethyl amine as Lewis acid and base, respectively. The yield was increased from 50% to 59% of the desired "Butadiene",

However, by switching dehydration agent the reaction conditions were drastically changed due to very low reaction temperatures (-68°C) and longer reaction times – up to 2h. Temperature increase in this case leads to an increase in the reaction rates, but has the disadvantage of causing a significant decrease of the yield of the produced cis-isomer.

Conclusions and future perspectives

Acceleration of the dehydration reaction in the zuclopenthixol production was performed in a successful way. Hence, the reaction time was decreased from 2h to just 3 min, while obtaining the same composition of the products. However, a complicated downstream processing of this process led to the investigation of new synthetic routes which would increase the amount of desired cis-isomer. Therefore, instead of using Brønsted acids, suitable combinations of Lewis acids and bases were tested. The results show increased yields of the desired product, but further investigations in this direction are necessary. As a final step, establishment of a control strategy for the most suitable chemical process has been done as well.

(586i) Optimal Antisolvent Addition Profile in a Continuous Plug Flow Crystallizer Through Population Balance Based Multi-Objective Optimization

Majumder, A., Loughborough University
Ridder, B. J.,
Nagy, Z. K., Purdue University

In recent years, the interest in continuous processing in pharmaceutical manufacturing has increased significantly due to its ability to provide several benefits such as improvement in processing time, consistency and reduction in expenses. In this work, we present the simulation study for obtaining optimal antisolvent addition profile in a 4-segment continuous plug flow crystallizer (PFC) with the objectives to maximize the mean size and minimize the coefficient of variation (CV). The system studied here is the crystallization of flufenamic acid in ethanol-water (antisolvent) system with experimentally determined parameters from Alvarez and Myerson [1]. The antisolvent can be introduced at the inlet of each segment. A steady state population balance equation (PBE), coupled with the necessary mass balance equation, is used to describe the evolution of the crystal size distribution (CSD) along the crystallizer. An optimization problem has been formulated which is then solved using nondominated sorting genetic algorithm II (NSGA II) available in Matlab. Due to the competing nature of the two objectives considered, no single optimal point is found, rather a set of optimal points called Pareto optimal points are obtained. The results are compared with those when 1 to 4 equal additions are used by keeping the total amount of antisolvent the same. It is found that the optimal profile is different from the profiles that distribute the antisolvent equally in 1 to 4 segments and is able to increase the mean size of the crystal significantly while maintaining the similar CV.

[1] A. J. Alvarez, A. S. Myerson: *Continuous plug flow crystallization of pharmaceutical compounds* (2010), Cryst. Growth Des. 10 (5), 2219-2228.

(586j) Dynamic Modeling of Encrustation and Mitigation Strategies in a Continuous Plug Flow Crystallizer

Majumder, A., Loughborough University
Nagy, Z. K., Purdue University

Encrustation, also known as scaling or crystallization fouling, on the wall of plug flow crystallizers (PFC) can cause major operating and economic problems in pharmaceutical industries. These include increased energy consumptions due to increment of thermal resistance and pressure drop, and unplanned plant shut down due to clogging of the crystals originating from reduction of flow area. In this work, a mathematical model for predicting the dynamic behavior of a PFC undergoing encrustation is presented. This model, which is adapted from the fouling model in heat exchangers by Coletti and Macchietto [1] and Chanapai [2], seeks to predict the encrustation thickness by tracking the thermal resistance. In addition, a population balance model for describing the crystallization process in the PFC is coupled with the encrustation model in order to obtain the appropriate concentration profile along the PFC as well as product crystal size distribution (CSD). Based on this model, mitigation strategies are also proposed that include heating cycles to overcome the problem of encrustation.

[1] Coletti, F. and Macchietto, S. (2011). A Dynamic, Distributed Model of Shell-and-Tube Heat Exchangers Undergoing Crude Oil Fouling. *Industrial & Engineering Chemistry Research* **50**(8): 4515-4533.

[2] Chanapai, A. (2010). Modeling of crystallization fouling in shell-and-tube heat exchangers. Master's thesis, Department of Chemical Engineering and Chemical Technology, Imperial College London.

(586g) Process Optimisation of Chromatography Units Using First Principles Modelling Techniques

Green, A. J.,
Martin, E., University of Newcastle upon Tyne.
McCreath, G.,

Purification is an important stage in the production of therapeutic proteins. It removes impurities and ensures quality of the product, one of the main stages in purification is chromatography. Ion exchange chromatography separates molecules based on their charge; there are two types of ion exchange chromatography; anion and cation. To develop a new biopharmaceutical process there is a lot of experimentation required to investigate it; this can be both time consuming and costly. This project looks into developing modelling tools which can be used to speed up the development of new processes.

A model was created which uses a first principles approach, where the molecular properties and the thermodynamics of the system are used to characterise it. The model uses continuity equations and the Langmuir isotherm to describe the action of adsorption and provide a tool for showing the breakthrough curve of the target molecule.

The model created required minimal experimentation to determine model parameters, PreDictor plates are used to estimate values for variables such as the equilibrium dissociation constant and the maximum binding capacity at different operating conditions. For this preliminary study pH was maintained at 7.5 so that only resin volume was tested. For this project data was collected from both the flowthrough and the elution stages to determine the difference between the two stages.

Other parameters that were required such as the coefficient of diffusion, were determined using thermodynamics so that the values calculated would not be specific to the process under certain conditions but would be adaptable to change under different conditions and different proteins.

To validate the model operational considerations such as pH, conductivity and protein load were considered to see how the model coped with changes to these variables. The data generated also allowed for the empirical information to be used to improve the model.

To investigate the effects of these variables design of experiments was used to obtain the maximum information and be able to define a design space for the specific molecule. It is aimed that this model alongside subsequent models can be used to promote quality be design, in that a new process can be understood prior to experimentation.

(586k) Chemoenzymatic Synthesis of Heparin for a Safer Bioengineered Alternative

Bhaskar, U., Rensselaer Polytechnic Institute
Yang, B., Rensselaer Polytechnic Institute
Fu, L., Rensselaer Polytechnic Institute
Takieddin, M.,
Zhang, F., Rensselaer Polytechnic Institute
Mousa, S. A., Albany College of Pharmacy and Health Sciences
Liu, J., School of Pharmacy, University of North Carolina
Dordick, J. S., Rensselaer Polytechnic Institute
Linhardt, R. J., Rensselaer Polytechnic Institute

Heparin, the first biopolymeric drug, possesses a wide range of structural heterogeneity owing to its biosynthesis. Its diverse fine structure is further complicated by an animal tissue-based recovery, leading to considerable structural differences within commercial heparin active pharmaceutical ingredients (APIs). Serious concerns about control of livestock, the primary source of heparin, have been raised since 1990s following a series of incidents involving Bovine spongiform encephalopathy, viral infections and prion contamination. Lack of quality control during initial recovery stages led to adulteration of pharmaceutical heparin with oversulfated chondroitin sulfate (OSCS), resulting in an international crisis in 2008 associated with 100 deaths in US.

The inherent problems with the animal tissue-based heparin production have motivated us to develop a commercially feasible chemoenzymatic heparin preparation process. This is based on bacterial fermentation of *E. coli* K5 to generate a capsular polysaccharide heparin precursor, that is then chemically *N*-deacetylated and *N*-sulfonated. A series of six recombinant enzymes, derived from heparin biosynthetic pathway and expressed in *E. coli*, are then used to epimerize uronic acid residues and sulfonate the C2, C3 and C6 positions. These modifications are carried out in tandem with sensitive analytical techniques and bioassays to result in a series of controlled structural changes. Bioengineered heparin consists of 86% trisulfated disaccharide resembling the highly sulfated structure of heparin. The formation of 9.4% lyase resistant tetrasaccharides upon digestion with heparin lyase 2 is consistent with the presence of antithrombin binding pentasaccharide regions in the range of USP heparins (3.9-6.7%). 1D-^{1}H NMR and HMQC techniques were employed to elucidate presence of 3-*O*-sulfo and I_{2S} peak in anomeric regions. Peaks corresponding to critical features in the IdoA and GlcN residues, including 2-*O*-sulfo, *N*-sulfo, *N*-acetyl, 3-*O*-sulfo, and 6-*O*-sulfo, have been fully assigned by ^{1}H-NMR and HMQC spectra. Investigation of *in-vitro* anticoagulant activity confirms the biological activity of chemoenzymatically derived heparin with an anti-factor Xa activity of 205.8 U/mg and an anti-factor IIa activity of 225.2 U/mg (anti-factor Xa:anti-factor IIa ratio of 0.91) complying with USP requirements. These analyses showcase generic equivalence of bioengineered heparin to animal derived heparin API. A more robust process based on immobilized enzymes, enhancing enzymatic stability and increased reusability, is under development and will be employed to generate bioengineered heparin for *in-vivo* studies.

(586l) Implementation of a Novel Inline Particle Size Measurement Tool to Monitor Pellets Produced By Hot Melt Extrusion and Die Face Pelletizing

Treffer, D., Graz University of Technology
Wahl, P., Research Center Pharmaceutical Engineering GmbH
Hormann, T., Research Center Pharmaceutical Engineering GmbH
Markl, D., Research Center Pharmaceutical Engineering GmbH
Eitzlmayr, A., Graz University of Technology
Schrank, S., Graz University of Technology
Koscher, G., Research Center Pharmaceutical Engineering GmbH
Roblegg, E., University of Graz
Khinast, J. G., Graz University of Technology

Hot melt extrusion is a promising manufacturing opportunity to increase the water solubility of poor water soluble drugs. The increase is obtained by the formation of various different types of solid dispersion. The extruder produces homogeneous molten material, which is mostly processed to intermediate products such as pellets or powders. Pellets are often desired for the downstreaming into the final drug product, e.g. capsule filling, tablet compaction, injection molding, due to excellent flowability and content uniformity. Cold strand cutters are commonly used for their production. The aim of our work is the establishment of hot die face cutting as production technology in the pharmaceutical industry. The advantage of this technology is that almost spherical pellets can be obtained with suitable formulation and process parameters [2],[3]. The pellet formation process is a complex relation of rheology, cooling intensity, extrusion speed, die geometry, knife design and speed. Therefore, an inline particle monitoring system is a powerful tool for process development as well as monitoring of the production process.

The experimental setup comprises of the processing and the monitoring equipment. The processing equipment consists of a co-rotating twin screw extruder ZSK 18 (Coperion, Germany) equipped with KT20 loss in weight feeders (KTron, Switzerland) and a hot die-face pelletizer (Automatik Plastics Machinery GmbH, Germany). The novel Eyecon™ (Innopharma labs, Ireland) was used for inline monitoring of the particle properties. Reference particle analysis was performed with a QICPIC System (Sympatec, Germany).

The Eyecon™ camera is a method based on diffuse light and uses three different light sources, which are circular arranged around the lens. This special illumination methods allows to derive three dimensional information from a two dimensional image. This additional information is used for improved particle edge detection, especially by overlapping particles. In this work different implementation approaches to the pelletizing line are investigated. Encountered problems are discussed and solutions provided. The best implementation approach was chosen for a comparison with the reference method and showed good agreement. In addition, it was recognized that the color images are useful to obtain information about the surface morphology of the produced pellets.

References:

1. Breitenbach J.: Melt extrusion: from process to drug delivery technology, *European Journal of Pharmaceutics and Biopharmaceutics* 2002, 54, 107-117.
2. Roblegg E, et al.: Development of sustained-release lipophilic calcium stearate pellets via hot melt extrusion, *European Journal of Pharmaceutics and Biopharmaceutics 2011, 79, 635-645.*
3. Bailleck S; Rein H, Preparation of starch-based pellets by hot-melt extrusion, *European Journal of Pharmaceutics and Biopharmaceutics, 2011, 79, 440-448*

(586s) Characterization of Solid-State Forms and Phase Transformations of Clindamycin Phosphate

RAN, R., State-Key Laboratory of Chemical Engineering

Two previously unknown solvates (ethanol-water solvate I and methanol-water solvate V), one new hydrate, Form III, and three novel polymorph forms (polymorph II, IV, VI) of clindamycin phosphate are discovered and characterized by various analytical techniques such as powder X-ray diffraction, differential scanning calorimetry, and thermogravimetric analysis. Solvate V is structurally characterized by single-crystal X-ray diffraction. The combined experimental and computational study was undertaken to understand the relationship among the discovered solid-state forms. Polymorph VI is obtained by the desolvation of solvate V, the desolvation mechanism has two steps, the solvent molecules (one methanol and two water) escape, and the solute molecules rearrangement by the layered structure split thinner. Base on the study, the desolvation of solvate I to polymorph IV may resemble solvate V. Upon heating, polymorph IV transform to polymorph II by solid-state transition. By suspension transition and solubility determination experiment, solvate I, V, polymorph II, IV, VI can transform to Form III in a pure water solution. Meanwhile, Form III also can convert to solvate I, V in a mix ethanol-water or a mix methanol-water solution respectively. The conversion relationships among the six solid forms have been illustrated.

(586m) Soft-Sensors: Efficient PAT Tools for Bioprocess Development Along QbD Principles

Sagmeister, P., Vienna University of Technology

Patrick Sagmeister patrick.sagmeister@tuwien.ac.at

Patrick Wechselberger patrick.wechselberger@tuwien.ac.at

Mohammadhadi Jazini mohammadhadi.jazini@tuwien.ac.at

Andea Meitz andrea.meitz@tuwien.ac.at

Timo Langemann timo.langemann@tuwien.ac.at

Christoph Herwig christoph.herwig@tuwien.ac.at

The following abstract is an extension of the following contribution:

Sagmeister P, Wechselberger P, Jazini M, Meitz A, Langemann T, Herwig C. 2013. Soft sensor assisted dynamic bioprocess control: Efficient tools for bioprocess development. Chem. Eng. Sci. 96:190–198.

Introduction

Bioprocess development along Quality by Design (QbD) principles aims at the identification and understanding of interactions between process parameters and product quality and process performance attributes. Since bioprocess developing time is limited and the investigation and optimization of a bioprocess is a time, cost and labour intensive task, strategies are required to accelerate process development and reduce time-to market latency of biotechnological products.

Soft-Sensors are process analytical devices that grant access to important non-measured process variables by mathematical processing of readily available process data. While the potential of soft sensors for manufacturing is generally recognized and reflected by many excellent contributions, their potential for efficient bioprocess development and bioprocess design purposes has not been exploited so far.

The task of soft-sensing in process control and monitoring in process development differs conceptually from soft-sensor applications in manufacturing. In the latter case huge amounts of process data are typically available that can be used to construct first-principle based or statistical process models. In process development no or very little prior data on the strain or process is available and soft-sensing acts as a tool for the development of strain- and process knowledge. This conceptual difference discards artificial neural networks (ANNs) and multivariate soft-sensing approaches for the investigation of unexplored systems within process development. In contrast, elemental balancing approaches for bioprocesses (JobÃ© et al., 2003) are attractive for the design of a process development soft sensing strategy: no training data sets are required, simple measurable input parameters and variables are used, they can be checked for gross errors via reconciliation procedures (van der Heijden et al., 1994).

In this contribution we present what is to our knowledge the first soft-sensor based control strategy capable of effective (dynamic) control of specific substrate uptake rates that is independent of training data sets, elaborate strain specific information, fixed or known yield coefficients or offline measurements. Therefore, the first soft sensor control strategy that qualifies as process development tool is presented.

Materials and Methods

A soft-sensor on the basis of a redundant equation system involving the Degree of Reduction (DoR) and Carbon (C) balance was used for the estimation of the real-time biomass concentration and used for closed-loop control of specific uptake rates using a mass balance based control approach.

Results and Discussion

The power of the presented soft sensing approach for bioprocess development was demonstrated on microbial bioprocess development and bioprocess design tasks: (1) the control of the specific substrate uptake rates in a non-induced *E. coli* decellerostat culture aiming at the determination of the specific acetate production as a function of the specific substrate uptake, (2) the static control of the specific substrate uptake rate in an induced P. Pastoris AOX expression system, (3) the dynamic control of specific substrate uptake rates under induced conditions in a *E. coli* pET expression system and (4) the simultaneous dynamic control of the specific glucose and arabinose uptake rates in an *E. coli*pBAD expression system.

(1) Dynamic control of a decellerostat culture aiming at the investigation of specific acetate production as a function of the specific substrate uptake rate

Acetate as overflow metabolite in *E. coli* has reported negative effects on recombinant protein production as well as on RNA, lipid and protein synthesis and can therefore be considered critical for recombinant processes (Koh et al., 1992). In fed-batch processes, acetate formation can be avoided by limiting and controlling the specific substrate uptake rate through the fed-batch feeding profile. However, this demands the strain specific knowledge of the onset of acetate production, i.e., which is maximum specific substrate uptake rate $q_s$without the formation of acetate.

The real-time biomass concentration was estimated *via* the biomass estimation device .The feed rate set point was calculated in real-time using the estimated biomass concentration, the estimated broth volume as well as the constant feed concentration (400 g/l) and feed density (1100 g/l). Offline samples were taken every 30 min for the determination of biomass dry cell weight as well as acetate and used for the calculation of the specific acetate production rate q_{ace} as well as the specific substrate uptake rate q_s. An offset between the q_s calculated from the offline biomass dry-cell weight determination and the q_s set-point was detected, converging in within to small q_s values. The calculation of q_s/q_{ace} from offline data followed by linear regression allows the extraction of the strain specific function $q_{ace}f(q_s)$. The q_s threshold for acetate production can be read from the interception of this function with the *x*-axis.

The bioprocess developer can obtain the information how long it takes for acetate (accumulated in the batch phase) to be consumed within the fed-batch at a specific substrate uptake rate q_s.

(2) Static control of the specific substrate uptake rate in an induced *P. pastoris* AOX expression system

Because of the dependency of the productivity of horseradish peroxidase produced by the recombinant *P. pastoris* on substrate uptake rate, the control of a static *qs* set point is of great interest (Dietzsch et al., 2011). Here it is demonstrated, that the control of a static *qs*set point can be achieved without knowledge on known yield coefficients and the assumption of constant yield coefficients.

A batch phase and consequently a fed batch phase were conducted to achieve a high cell density culture ready for induction. Consequently, the presented control strategy was used for the control of the specific methanol uptake rate in the induction phase. The desired control of the specific substrate uptake rate was achieved without knowledge of the yield coefficient with a maximum deviation of 10%.

(3) Dynamic control of the specific substrate uptake rate in an induced *E. coli* C41 (pBMPPET) expression system

The control of dynamically changing setpoints of the specific substrate uptake rate qs is considered of practical interest for bioprocess development, especially for the design of feeding strategies for recombinant processes (Dietzsch et al., 2011). However, dynamically changing yield coefficients are frequently observed in induced conditions, restricting the applicability of feed-forward control strategies as well as control strategies based on total carbon dioxide evolution or total oxygen consumption.

Here, the applicability of the presented soft-sensor control strategy for the dynamic control of the specific substrate uptake rate qs on an induced process with dynamically changing yield coefficients due to the effect of recombinant protein production is demonstrated.

(4) Simultaneous dynamic control of the specific substrate and inducer uptake rates in an *E. coli* expression system aiming at the investigation of mixed feed metabolic capabilities

Aiming at the investigation of mixed feed capabilities of an *E. coli* C41 strain carrying the plasmid pBK-BMP, the culture was subjected to two independent C sources (glucose as main C-source, arabinose as second C-source and inducer for recombinant protein production) following the *qs*control strategy. The goal of the dynamic experiment was to investigate the capability of l-arabinose metabolization in a range of 0 to 0.3 g/g/h specific uptake of l-arabinose in the presence of an uptake of d-glucose of -0.2 g/g/h.

The *qs* glucose setpoint was kept constant at a level of −0.2 g/g/h and the *qs*arabinose was subjected to a linear ramp starting from -0.05 g/g/h to −0.3 g/g/h. No l-arabinose was found to accumulate, demonstrating that the culture is able to metabolize l-arabinose in a flux range from −0.05 g/g/h to −0.3 g/g/h in the presence of specific glucose uptake of −0.2 g/g/h. Obtained specific substrate uptake rates were calculated on the basis of biomass dry cell weight measurements. A deviation of 3 % was observed.

Conclusion

In this contribution the dynamic controllability of multiple specific substrate uptake rates using a first-principle soft sensor independent of known or dynamically changing yield coefficients in recombinant bioprocesses is demonstrated for the first time. The presented control strategy holds potential to become a key process analytical technology (PAT) tool for bioprocess development; hence a tool for efficient design, analyzing and control of manufacturing processes.

References

Dietzsch C, Spadiut O, Herwig C. 2011. A fast approach to determine a fed batch feeding profile for recombinant Pichia pastoris strains. *Microb. Cell Factories* **10**:85.

Van der Heijden RTJM, Heijnen JJ, Hellinga C, Romein B, Luyben KCAM. 1994. Linear constraint relations in biochemical reaction systems: I. Classification of the calculability and the balanceability of conversion rates. *Biotechnol. Bioeng.* **43**:3–10.

JobÃ© AM, Herwig C, Surzyn M, Walker B, Marison I, von Stockar U. 2003. Generally applicable fed batch culture concept based on the detection of metabolic state by online balancing. *Biotechnol. Bioeng.* **82**:627–639.

Koh BT, Nakashimada U, Pfeiffer M, Yap MGS. 1992. Comparison of acetate inhibition on growth of host and recombinantE. coli K12 strains. *Biotechnol. Lett.* **14**:1115–1118.

(586ag) Evaluation of Film Coated Tablet Surface Roughness Via a Latent Variable Design-of-Experiments (DOE)

Miesle, J. E., Eli Lilly & Company
Manley, L., Eli Lilly & Company

A one-third fractional factorial design of experiments (DOE) was created utilizing latent variables for two separate spray bars in order to evaluate the impact of those parameters on coated tablet appearance. The variables evaluated were environmental equivalency factor (EEF), spray droplet size and coating pan exhaust temperature while utilizing either a Thomas Engineering Spectrum spray bar or a Spraying Systems VMAU spray bar in a 24" Thomas CompuLab coating pan. EEF values were calculated using the Thomas Engineering TAAC program. Droplet sizes were calculated by using an empirical model developed internally from data collected during characterization trials at Spraying Systems, Inc.

Placebo tablets (10 mm round standard concave, debossed on one face) were coated per the DOE conditions and then evaluated for gloss readings and surface roughness. Gloss readings were measured by Colorcon and represent the amount of light reflected by the tablet. Surface roughness was measured by non-contact three dimensional (3D) profilometry via a Keyence VK-X100 3D laser scanning microscope. Roughness measurements were recorded on both faces of five tablets from each DOE run. Data filtering and smoothing algorithms were applied to each measurement through the use of a template and results were tabulated. Coated tablet surface roughness values were found to be highly correlated with the Sauter mean droplet diameter (D[3,2]) and to a lesser extent with EEF.

(586af) Model Based Robustness Analysis of An Industrial Ion Exchange Step

Hansen, T. B., Novo Nordisk A/S
Hansen, E. B., Novo Nordisk

The application of mechanistic models for process development and optimization has been a topic of discussion in the Pharmaceutical industry for many years. Mechanistic models are considered as an alternative to a standard DoE type empiric approach. Most often the DoE approach is preferred based on a cost-benefit evaluation where the time consumption and the complexity are found to extensive when using a mechanistic modelling. In our company we use mechanistic models for the development of selected downstream process steps. In the purification of peptides and proteins it is of great advantage to have the ability to predict the behaviour of the purification steps i.e. retention of the product and various closely related impurities as a function of variables like salt concentration, gradient and loading conditions. The ability to simulate the chromatographic process facilitates fast development and optimization, and it assists robustness analysis of the purification step. The present work demonstrates how simulation, using a mechanistic model of ion exchange chromatography, can be applied for robustness analysis of a process step.

(586x) Rate-Based Information Processing Approaches for Bioprocess Development Along QbD Principles

Sagmeister, P., Vienna University of Technology

Patrick Sagmeister patrick.sagmeister@tuwien.ac.at

Patrick Wechselberger patrick.wechselberger@tuwien.ac.at

Christoph Herwig christoph.herwig@tuwien.ac.at

Details on the topics presented can be found in following contributions:

Sagmeister P, Wechselberger P, Herwig C. 2012. Information Processing: Rate-Based Investigation of Cell Physiological Changes along Design Space Development. *Pda J. Pharm. Sci. Technol.* **66**:526–541.

Sagmeister P, Wechselberger P, Jazini M, Meitz A, Langemann T. 2013. Soft sensor assisted dynamic bioprocess control: Efficient tools for bioprocess development. *Chemical Engineering Science.* **96:** 190-198.

Wechselberger P, Sagmeister P, Herwig C. 2012. Real-time estimation of biomass and specific growth rate in physiologically variable recombinant fed-batch processes. *Bioprocess Biosyst. Eng. (online first)*

Wechselberger P, Sagmeister P, Herwig C. 2013. Model-based analysis on the extractability of information from data in dynamic fed-batch experiments. *Biotechnol. Prog.* **29**:285–296.

Abstract

Regulatory initiatives summarized under the term quality by design (QbD) promote science and risk-based pharmaceutical bioprocess development strategies. During upstream process development as well as manufacturing a plethora of on- and off-line data is collected but typically not efficiently exploited and used for the demonstration of process understanding. Here we want to present recent developments in the field of data processing (quality check and condense your data into information) as well as multivariate information processing (extract knowledge from your information) of fermentation data. On the basis of the presented concepts a novel roadmap for data- and information processing of fermentation data is presented.

1) Soft sensing- Real-time sensing of physiological information

Soft sensors are process analytical devices that use readily available on-line signals to calculate important process descriptors and hence grant on-line information on the state of the process. Next to on-line applications, first principle soft sensors can be run on historical data sets to extract non-measured variables such as the biomass concentration, specific rates or yield coefficients that stand in connection to cell physiology. Applications, limits and practical aspects of the use of soft-sensors for on- and offline use are demonstrated.

2) Information quality control: Reconciliation as a statistical tool to ensure bioprocess information quality

Data Reconciliation refer to statistical approaches to remove measurement error out of data through the application of defined constrains. Little prior knowledge is required and the signal to noise ration of the extracted information can be greatly enhanced. Furthermore, a statistical test can be applied to detect gross measurement errors. This is especially useful for further information processing approaches (e.g. multivariate data analysis), since outliners can be efficiently detected and removed on a sound scientific basis. Applications, limits and practical aspects of the use of data reconciliation for data quality control are discussed.

3) Information Processing: Multivariate rate-based investigation of cell physiology

The application of multivariate techniques on fermentation data is still a challenging task. The use of processed data (specific rates and yield coefficients) instead of raw data for multivariate data analysis is highly promising, since variables have a direct physiological interpretation and the number of variables can be reduced prior to multivariate data analysis. The general concept of rate based multivariate data analysis is presented and applications are discussed.

4) A rate based data to knowledge roadmap for upstream bioprocess development

The tools mentioned above are combined to a novel general concept for the a) mining of data and information, b) rate based processing, c) statistical assessment as well as d) multivariate processing of information. The presented roadmap for process knowledge extraction is complementary to process development and continuous improvement along QbD principles.

(586n) In-Line Monitoring of a Fluid-Bed Coating Process With Optical Coherence Tomography

Sacher, S., Research Center Pharmaceutical Engineering GmbH
Markl, D., Research Center Pharmaceutical Engineering GmbH
Hannesschlager, G., Research Center for Non Destructive Testing GmbH
Leitner, M., RECENDT â€“ Research Center for Non Destructive Testing GmbH
Khinast, J. G., Graz University of Technology

In-line monitoring of a fluid-bed coating process with optical coherence tomography

Stephan Sacher[1], Daniel Markl[1], Günther Hannesschläger[2], Michael Leitner[2], Johannes G. Khinast[3]

[1] Research Center Pharmaceutical Engineering GmbH, Inffeldgasse 13/2, 8010 Graz, Austria

[2] Research Center for Non-Destructive Testing GmbH, Science Park 2, Altenberger Strasse 69, 4040 Linz, Austria

[3] Institute for Process and Particle Engineering, Graz University of Technology, Inffeldgasse 13/3, 8010 Graz Austria

Coating of tablets, pellets and granules is one of the most common unit operations in pharmaceutical industry. Uniformity and thickness of the coating layer are crucial for efficacy as well as for compliance. Therefore, not only because of different initiatives it is essential to monitor and control the coating process in-line.

Optical coherence tomography (OCT) features a contact-free non-destructive technique based on white light interferometry offering high resolution images of internal structures of solid materials. Independent on calibration models like e.g. spectroscopic methods, and small in space compared to terahertz, it represents an alternative process monitoring technique enabling to investigate coating thickness as well as coating homogeneity. Due to the fast acquisition rate it is even possible to measure single tablet or pellet parameters within a bulk of moving tablets.

In this study the applicability of OCT to monitor growth of coating in-line is validated. Therefore, an OCT system is integrated into a commercially available pilot-scale fluid-bed system. In a first step fluidized pellets with and without prior applied coating passing with high speed are measured. In addition in-line experiments with real-time monitoring of coating growth are performed. Results show that OCT is a suitable technique to measure coating thickness of even fast moving particles within pharmaceutical production processes.

(586ah) Novel Approach for Stabilizing Physical Form of a Drug Substance in Tablet Formulations

Nunes, C., Bristol Myers Squibb
Bindra, D., Bristol Myers Squibb
Gokhale, M., Bristol Myers Squibb
Thakur, A., Bristol Myers Squibb

PURPOSE

The objective was to develop a commercially viable tablet dosage form for a BCS class II oncology compound. The key challenge, in addition to low solubility (<1 ug/ml), was the tendency of the anhydrous drug substance to readily convert to a hydrate (a_w 0.26 - 0.36). The formation of the lower solubility hydrate form during shelf-life led to a 2-3 fold decrease in the oral bioavailability. The specific aim was to kinetically stabilize the anhydrous form of the drug in the tablets and achieve superior bioavailability.

METHODS

The kinetics of the anhydrate → hydrate transition were mapped for aqueous systems containing a series of polymeric excipients. Tablets were manufactured using the dry granulation and the wet granulation approaches. Further, the choice of binder and the mode of binder addition (dry, solution, or foam) were evaluated for the wet granulation process. Particle size effect (0.3 - 70 um) on bioavailability was also evaluated.

RESULTS

The anhydrous form of the API when suspended in aqueous medium converted to the hydrate in <24 hrs. Aqueous solutions containing cellulosic polymers such as hydroxypropyl cellulose (HPC, 0.1 - 5%w/w) successfully inhibited the form conversion. The kinetically stabilized suspensions showed a 2-fold improvement in bioavailability as compared to particle size reduction approach, emphasizing the importance of form control. The tablets containing HPC manufactured by the dry granulation process underwent rapid form conversion under high humidity conditions, attributable to ineffective distribution of HPC in the formulation. In contrast, the tablets prepared using wet granulation process using aqueous HPC solution as a binder showed a pronounced improvement in physical stability. The stability was further enhanced when HPC was added in as foam during granulation. The foam technology (Dow) led to a more uniform distribution of the stabilizer within the tablet matrix. The resultant tablets were stable under stressed conditions (40C/75%rh) for >3 months, and provided significantly superior bioavailability in humans as compared to formulations containing the hydrate form.

CONCLUSION

The API form conversion in drug product was successfully inhibited by incorporating a polymeric stabilizer in the formulation matrix. Generally, a dry granulation unit operation is utilized for processing of materials that are sensitive to moisture. This work demonstrated that effective distribution of the polymeric stabilizer in the API microenvironment via the aqueous foam granulation approach provided a drug product with superior physical stability. The approach enabled manufacture of tablets having enhanced bioavailability and extended shelf-life.

(586o) Design of An Efficient Control System for Flexible Continuous Tablet Manufacturing Process

Singh, R., Rutgers University
Sahay, A., Rutgers University
Ierapetritou, M., Rutgers University
Ramachandran, R., Rutgers University

Monitoring and close loop control is of great important in continuous manufacturing of pharmaceutical tablet to guarantee the end product quality. In the work reported here, an efficient control system has been designed for a flexible multipurpose continuous tablet manufacturing process. The flexible continuous tablet manufacturing process consists of direct compaction (DC) [1], wet granulation (WG) and dry granulation (DG) [2] routes where one route can be selected based on feed powder characteristics and the desired end product quality specifications. DC is the simplest route used for easily flowing material, the WG route can be used to improve the product quality, and the DG route can be used in place of WG if the API and/or Excipient are water sensitive. Three control strategies (PID, Model Predictive Control (MPC), and, hybrid MPC-PID) have been designed for this process and their performance regarding set point tracking and disturbance rejection has been compared using a flowsheet model simulated in gPROMS (PSE) and control interface DeltaV (Emersion).

The control systems have been implemented in the control studio of the DeltaV system and the inputs and outputs of the controller are connected with the process model simulated in gPROMS. IGEAR software and the gORUN feature of the gPROMS have been used to communicate between gPROMS and DeltaV system. The model is used in place of the plant to generate the data needed for design and performance evaluation of the control system. In direct compaction process the API composition, Relative Standard Deviation (RSD) and flow rate are controlled at blender, and the tablet weight and hardness are controlled at the tablet press. The selected control strategy is then integrated with the pilot plant through control hardware, software and sensors implementation.

NIR is used to monitor the API composition at blender outlet, CAMO Unscrambler X is used to develop the prediction model, Unscrambler Process Pulse uses Prediction engine and the prediction model to generate the composition signal from the measured spectrum. This signal is then communicated to the DeltaV control studio through MATLAB OPC communication protocol. Based on the API composition measurement, the controller implemented in DeltaV calculates the actuator setting (feeders rotation speeds) which is send to the plant through DeviceNet. Similarly, RSD a measure of blend uniformity is controlled using NIR as the monitoring tool and blender rotational speed as the actuator. The tablet weight is controlled through a cascade control scheme where the master controller generates the set point of pre-compression force, which is then tracked by a slave controller by manipulating the lower punch displacement of the tablet press. The lower punch displacement changes the amount of powder filled in the die. The tablet hardness is also controlled through a cascade control scheme where the master controller generates the set point for main compression force which is tracked by a slave controller by manipulating the upper punch displacement. The pre-compression and the main-compression forces are monitored through strain gauge (sensors inbuilt in the tablet press) while the tablet weight is monitored through a load cell/weighing instrument and NIR is used for hardness measurement. Alternatively, soft sensing approach can be also employed for weight and hardness measurements.

The objective of this presentation is two-fold: first to demonstrate the model-based design and performance evaluation of the different control strategies for flexible multipurpose continuous tablet manufacturing process and then to present the implementation of the best control strategy into the direct compaction line.

References

1. Singh, R., Ierapetritou, M., Ramachandran, R. (2013). European Journal of Pharmaceutics and Biopharmaceutics, http://dx.doi.org/10.1016/j.ejpb.2013.02.019.

2. Singh, R., Ierapetritou, M., Ramachandran, R. (2012). International Journal of Pharmaceutics, 438 (1-2), 307-326.

(586ai) Scale Up of the Chromatographic Process

Kidal, S., Novo Nordisk A/S
Hansen, E. B., Novo Nordisk

Scale Up of the Chromatographic Process

Steffen Kidal, Novo Nordisk A/S

The transfer of chromatographic processes from lab. scale to pilot and production scale is an important and challenging task in the pharmaceutical industry, and has been dealt with in several - more or less obvious - ways. Traditionally the process is scaled by increasing the column diameter while keeping the bed height, the velocity, and the volumes of the different phases (measured in column volumes) constant. Our approach was to find a practical flexible and consistent method, a tool, which requires only a few experiments to produce a safe process transfer.

The process transfer is considered as a two-step operation. The first step is the scale up of the column process, which follows the principle of scaling on a volume basis (constant residence time for the proteins). This concept is quite simple and does not involve any new theory. The second step deals with the extra column effects created by instrumentation (dead volumes, mixer characteristics etc.).

(586t) Continuous Direct Compaction As a Manufacturing Route for Highly Segregating Blends

Oka, S., Rutgers, The State University of New Jersey
Sahay, A., Rutgers University
Muzzio, F., Rutgers University

Pharmaceutical solid oral dosage formulations in which the difference in particle size of the active ingredient and excipients is large have been processed via continuous direct compaction to produce tablets that comply with quality regulations.

Granular mixtures with distinctly bimodal size distributions tend to de-mix. Granular segregation based on particle size is a bane in the pharmaceutical industry since a difference in the particle size of the active ingredient and the excipients can cause in-process demixing leading to poor content uniformity at the tablet level. Such formulations are thus granulated to ensure homogeneity at the powder level which translates into tablets with robust content uniformity.

Recent advancement in continuous pharmaceutical processing was leveraged and highly segregating granular mixtures were processed in a *continuous* direct compaction line to produce acceptable tablets. The phenomenon has been exhibited by performing trials with a variety of such formulations. The most extreme case is a formulation where the particle size of active ingredient is 12 times that of the major excipient. The authors believe that the fundamental premise of a continuous process, that, at steady the rate of input equals the rate of output coupled with the fact in a continuous process, a much lower volume of material is present in the system as opposed to a batch process where the whole batch is 'in process' are key reasons behind being able to handle such systems via direct compaction.

Direct compaction is a cheaper processing route than wet or dry granulation and involves fewer unit operations. Continuous direct compaction eliminates uncertainty associated with its traditional batch cousin and thus presents the possibility of making several products via this route which are currently being made via the granulation routines. This could prove to be an industry game changer.

(586aj) Development of a Robust Film Coating Process: The Impact of Wetting Dynamics On the Film Stress - Film Adhesion Balance

Sun, J., Merck & Co., Inc.
Lu, J., Merck & Co., Inc.
Conway, S. L., Merck & Co., Inc.
Seymour, E., Merck Sharp & Dohme Corp.
Eickhoff, W. M., Merck & Co., Inc

During development of a tablet film coating process, a film coating defect was observed which obscured tablet debossed markings, rendering the identifying markings to be illegible and resulting in batch failure. Defect analysis revealed that the defects were associated with film coat bridging across the debossed features. Literature indicates that this type of coating defect occurs when the internal stresses in the film coat exceed adhesion forces between the film coat and core tablet substrate. It was hypothesized that tension in the film coat was exacerbated during drying when the film coat contracts and is relieved when the coat and pulls away from the core tablet; supporting evidence was obtained. Characterization studies evaluating the wetting dynamics of the film coat suspension on the core tablet demonstrated that coating droplet extent of spreading and penetration on the core tablet surface was lower than with other formulations which did not exhibit the bridging defect. Increased film thickness and rate of coating deposition also increased the bridging tendency. A similar film coating defect in which the film was seen to peel away from the core tablet surface is attributed to the same mechanism as the bridging defect – i.e., film coating internal stresses which are higher than the coating-to-core adhesion forces. Modifications to the film coating process to balance these stresses were focused on process parameter adjustments to influence drying rate and coating uniformity. Suspension solids concentration was highlighted as an additional process factor which could potentially minimize the risk of tablet appearance failure. A 2^3 design of experiments was conducted evaluating film coating process factors. Regression analysis indicated that the inlet air flow rate was the factor which most strongly influenced the bridging defect, and the ratio of air flow rate to spray rate was the factor which most strongly influenced the peeling defect. A relatively narrow but robust film coating process design space was demonstrated for this product at both the pilot and commercial scales.

(586u) Design of An Interface for Integrating PAT Tools in Continuous Manufacturing for in-Line Real Time Sensing

Sahay, A., Rutgers University
Jayjock, E., Rutgers University
Sanchez, E., Janssen Pharmaseuticals
Romanach, R. J., University of Puerto Rico at Mayaguez
Muzzio, F. J., Rutgers, The State University of New Jersey

Currently, pharmaceutical manufacturing is a batch driven process with unit operations such as granulation, milling, coating etc. This approach has a number of disadvantages such as big footprint of equipment, higher costs and lower product quality. There is a push within the industry to adopt continuous manufacturing with inline testing and control. Such systems will have much smaller equipment requirements such that the whole line could be built in one space. They will be modular so as to be able to incorporate different processes, such as wet/dry granulation with relative ease. Also, the incorporation of inline testing will lead to better product quality and reduced wastage.

Under the PAT(Process Analytical Technology) framework proposed by the FDA there is a strong push to developing sensing techniques to make the manufacturing process more robust and incorporate Quality by Design. A majority of the commercially available analytical instruments are bench top in nature and need to be adapted for real time sensing. One of the challenges lies in building the methodology for real time sensing and sampling Here we describe the design and selection of a chute for attachment of NIR instruments to a continuous manufacturing line. This setup can be used for in line detection of blend uniformity. Furthermore it can be used as a control variable for closed loop control and process optimization.

(586y) Near-Infrared Spectroscopy for Predicting Dry Granulated Ribbon Density: Effect of Modeling Approach, Chemical Composition, and Spectroscopic Signature On Calibration Model's Performance

Talwar, S., Duquesne University
Nunes, C., Bristol Myers Squibb
Stevens, T., Bristol Myers Squibb
Nesarikar, V., Bristol Myers Squibb
Timmins, P., Bristol Myers Squibb
Anderson, C. A., Duquesne University
Drennen, J. K. III, Duquesne University

The powder properties of pharmaceutical materials play a critical role in determining the manufacturability and performance of solid dosage forms. For a product manufactured by dry granulation, a commonly utilized unit operation in pharmaceutical industry; the properties of the resultant granules can have a direct impact on the downstream operations (milling, tabletting) as well as on product performance (disintegration, dissolution and friability). Dry granulation is often accomplished by roller compaction process; wherein the density of the output ribbons is often a critical quality attribute (CQA) and can impact the particle size of milled granules, blend flowability and tabletability. Defining and maintaining optimal ribbon density is therefore one of the key objectives during drug product process development and scale-up.

Our research objective was to develop an objective *NIR based spectroscopic method for on/at-line determination of ribbon density*. The raw NIR spectrum of a sample displays information pertaining to its chemical composition, as well as the three-dimensional physical structure. Thus, it was pertinent to investigate the impact of chemical variability on the spectral representation of a physical attribute, i.e. density. Additionally, with respect to the NIR calibration model, the choice of modeling algorithm can influence model's specificity to the property of interest, and its accuracy. Thus, the specific aims were to evaluate the impact of (i) *chemical composition* on spectroscopic signatures, and (ii) *mathematical algorithms* on the NIR model performance.

The calibration model was developed utilizing standardized compacts of excipient based formulations comprising of diluents/binders (microcrystalline cellulose + lactose), disintegrant (0-3%), glidant (0-1.5%) and lubricant (0-1.5%). The model was then tested on roller compacted ribbons and compacts containing an active pharmaceutical ingredient (API) at various loads (0-20 % w/w). The standard compacts were prepared using a compaction emulator (Stylcam^{TM}) and characterized using AOTF-NIR spectrometer. Key formulation and process variables evaluated were: grade of excipients, ratio of diluents, levels of lubricant and glidant, and compression pressure. The densities of standard compacts were also determined using volumetric techniques (GeoPyc, AcuPyc and dimensional measurements). Two modeling approaches were evaluated used for density calibration: 1) a univariate model between density and spectral slope, and 2) multiple linear regression (MLR) approach which modeled density as a function of spectral slope augmented with formulation information.

Near-infrared (NIR) spectroscopy is a powerful technique for characterization of the physical attributes of pharmaceutical samples, in addition to its utility in chemical analysis. The physical integrity of a sample affects its absorption of NIR radiations. An increase in sample density produces a lesser proportion of air-material interfaces, in comparison to material-material interfaces. This suggests a lower probability of NIR photons encountering a change in refractive index, which causes radiation scattering. This results in a net increase in absorption, depicted as a positive baseline shift in a sample's NIR spectrum. Consequently, the slope of the baseline referred to as ?*spectral slope*? can be used to model for sample density via an univariate modeling approach. NIR analysis of the compact and ribbon density samples revealed that the main effect of chemical design variability on spectroscopic signature of density was a decrease in spectral slope values. An increase in API concentration in the ribbons and compacts produced smaller slopes even at higher compression forces. The reasons were attributed to the appearance of new shapes/chemical peaks in the raw spectrum. Additionally, due to the poor deformation behavior of the API, an increase in drug loads produced compacts of lower densities with smaller spectral slopes.

As a next step, the NIR data in conjunction with the formulation and process related variables was used to build a model based on a MLR approach. The data analysis indicated that the MLR model (error <1%) statistically outperformed the spectral slope based univariate model. Further, the model also enabled prediction of ribbon density in presence of API, with >95% accuracy for 1-10% drug load. Addition of formulation information in the MLR model provided a better description of ribbon density; which led to an improved model performance compared to spectral slope approach.

The use of NIR for ribbon density measurement has several benefits, e.g. non-destructive, rapid, improved statistics, potential for at/in-line monitoring, over traditional volumetric methods. The NIR method showed capability for prediction of ribbon density at pilot-scale, using calibrations developed on compacts created at small scale. The study also demonstrated that a calibration model based on standard compacts can effectively be utilized to predict density of ribbons samples obtained from product manufacturing operations. The MLR model approach has a wide-ranging applicability during development of low dose formulations, as the model is largely unaffected by the variations in API or excipients encountered during formulation robustness and process development stages.

(586z) Virtual High-Throughput Screening By Multi-Dimensional Characterization Techniques

Herring, R. III, Auburn University
Eden, M. R., Auburn University

The design and selection of chemicals with targeted properties and activities has greatly benefited from the application of quantitative structure activity relationships (QSAR's). This approach involves the regression of various molecular descriptors, such that a property or activity of interest is accurately modeled as a function of the underlying molecular features. The advent of powerful variable selection techniques has allowed for these models to become highly predictive in nature, which can be taken advantage of in the search for optimal candidates meeting certain predetermined property/activity requirements. It has been shown that each property of interest is best characterized by it's own unique combination of molecular descriptors. These descriptors capture widely varying aspects of a molecule's features and, among many ways, can be differentiated by the dimensionality of information they capture. Simple, low-dimensional descriptors such as molecular weight or number of aromatic rings, while easy to calculate, cannot explain more complicated phenomena such as ligand receptor interactions. On the other hand descriptors containing information such as spatial electrostatic character can explain these interactions, but are much more computationally intensive and vulnerable to statistical noise. The use of these QSAR's in a predictive manner towards identifying promising candidate molecules in a large region of chemical space is known as virtual high-throughput screening (VHTS). Used along-side conventional laboratory based HTS methods, VHTS can significantly reduce the time and costs associated with developing novel chemicals and drugs meeting certain property and activity requirements. The inclusion of descriptors with higher dimensionality, specifically 3D descriptors, has proven very successful but still has its limitations. Most often, these descriptors are calculated from the local energy minima of each molecule in a data set, which is not highly representative of the actual conformations found in solution. In response, techniques have been developed to calculate these spatial descriptors from a conformational ensemble, which can include information such as a molecules flexibility and is commonly known as 4D-QSAR. The utilization of these models in a predictive manner can become quite computationally intensive, especially when searching large regions of chemical space, thus necessitating methods to expedite the search. This contribution details an algorithm useful for VHTS in large regions of chemical space with descriptors of varying dimensionality (0D-4D). The approach utilizes concepts in graph theory and employs genetic algorithm to enable the consideration of a large search space subject to multiple property constraints. The signature molecular descriptor is used to compress the molecular information in a canonical form, which reduces memory storage requirements and facilitates an efficient combinatorial search. A case study will be presented to exemplify the applicability of such an approach.

(586ak) An Alternate Coating Method for Tablets: Wurster Technology

Stamato, H., Bristol-Myers Squibb
Pafiakis, S., Bristol-Myers Squibb
Keluskar, R.,
Levins, C.,
Martin, K., Bristol-Myers Squibb
Lin, S.,
Chen, W., Bristol-Myers Squibb Company
Fiske, J., Bristol-Myers Squibb
Tejwani, R.,
Timmins, P., Bristol Myers Squibb

An Alternate Coating Method for Tablets: Wurster Technology

Tablets are routinely coated for appearance, controlled release and in some cases with active ingredients. While pan coating is an ancient process which has been updated through the use of perforated pan technology to meet modern pharmaceutical processing; air suspension coating, such as that accomplished in a Wurster device has rarely been applied. While the mechanical forces which a tablet could experience are of concern and are most often cited as the cause for discounting the Wurster process for tablets, the proper selection and setup of equipment and design of tablet properties can assure a successful coating process.

Although the basic coating technology - applying an atomized liquid to moving tablets - is the same for each of the techniques, the airflow, tablet flow, and contact with the atomized liquid is different for the Wurster process and can actually impart some advantages. For example the droplet flight to the tablet is typically shorter for a Wurster compared to a coating pan, which can reduce the amount of spray dried coating generated during the process. The contact between liquid and tablet in a Wurster occurs when the tablets are spatially separated in lean phase pneumatic transport upwards in the central column, rather than the coating pan flow of tumbling against each other in a packed bed, which widens the range of acceptable coating parameters. Unlike tablets tumbling in a pan the material flow in the Wurster device has a definite pattern which means it can approximate an idealized plug flow reactor rather than a continuous stirred tank.

Experiments have shown that small scale Wurster devices can be used to prepare small samples of product quickly allowing the screening of prototypes in early development while consuming a minimal amount of material. The selection of equipment and conditions for coating tablets in a Wurster and the required tablet properties to obtain successful coating results will be discussed. Tablets of up to 1200 mg weight were coated without compromising the quality of the product. A comparison of tablets coated using pan and Wurster technology at similar scales will also be discussed.

(586p) Optimization of Protein Pegylation

Shang, X., McMaster University
Corbett, B., McMaster University
Mhaskar, P., McMaster University
Ghosh, R., McMaster University

PEGylation is a generic term for chemical conjugation of polyethylene glycol (or PEG) to biomacromolecules such as proteins. PEGylation is frequently used to improve the properties of therapeutic proteins. While this type of chemical reaction is easy to carry out, it typically results in the synthesis of complex mixtures of different PEGylated forms, i.e. mono-, di-, tri-, and higher-PEGylated protein. With many PEGylated therapeutic protein products the mono-PEGylated protein is usually the desired form. Formation of by-products in a PEGylation reaction represents material loss as well as the inconvenience of a challenging purification process that would follow. The problem of low selectivity of mono-PEGylation could be addressed through the use of specific reactions such as N-terminal PEGylation. However, even with such specific reactions, significant amounts of by-products are typically synthesized. Generally speaking, a trade-off between selectivity and yield of desired product is observed in most PEGylation reactions. We present a systematic attempt at optimization of batch protein PEGylation. To this end, initially a first principles model (with the associated reaction network) is presented with kinetic rate constants determined using the experimental data. The dynamic model captures the effect of variation in initial condition as well as the pH at which the experiment is run. Having established the validity of the model beyond the fit data by additional experiments, the model is then utilized to determine the optimal pH, duration and initial recipe that maximize the formation of the mono-PEGylated product. The benefits of operating at the conditions suggested by the optimization problem are finally demonstrated though additional experiments.

(586am) Feasibility of Spray Dried Dispersion (SDD) Based Low Dose Tablet Formulation

Ferrie, P., Bristol Myers Squibb

Nunes, C., Bristol Myers Squibb

Reddy, J. P.,

Dennis, A.,

Banks, E.,

Wray, P.,

Gamble, J., Bristol-Myers Squibb

Stamato, H., Bristol-Myers Squibb

Timmins, P., Bristol Myers Squibb

BACKGROUND

The design and development of low-dose solid drug products presents a unique set of challenges during drug product development. The key concern with a low-dose product is achieving and maintaining acceptable content uniformity of the therapeutic active. In addition, frequently there is a need to utilize enabling technologies to increase bioavailability of poorly soluble drugs. Rendering the active therapeutic moiety to be in an ?amorphous state' is a promising approach to obtain a material with high kinetic solubility. In pharmaceutical industry, the bulk amorphous dispersions are often manufactured using the spray drying technology. The process involves dissolving the active drug and polymeric carrier(s) in a solvent system, followed by evaporating the solvent in a spray dryer. Notably, the spray dried materials often have physical properties (size, shape and flow) that are different than those of traditional solid dose excipients.

This work assesses the feasibility of developing a low dose SDD formulation with a representative material, composed of 30% API and 70% hydoxypropyl-methylcellulose acetyl succinate (HPMC-AS) polymer. To support clinical development a wide dose range is often necessary; it is preferable to use a common SDD (same drug loading) to ensure consistent performance and to minimize complexity with regard to SDD process development. Hence the provision of low dose tablets, for example as low as 0.1mg, using very small quantities of SDD, necessitates high assurance of content uniformity. The manufacturing process described in this work follows a dry granulation process which was selected to provide for acceptable physical stability, and to reduce the risk of segregation and poor content uniformity.

Traditional models for predicting content uniformity (Rohrs *et al.*) based on mean particle size and distribution skew suggest that the risk of content uniformity failure for typical SDD material (e.g. D50 40μm) in a 1mg formulation is low ($P<0.1\%$), whilst for a 0.1mg tablet the risk of failure rises particularly as the size distribution becomes skewed, for D90/D50 ratio >2, P failure >8%. The feasibility of using low dose SDD based formulations therefore presents unique challenges which to date have not been fully explored.

METHODS

Tablets were manufactured at a 10 kg scale (0.1 mg and 1.0 mg potency) using 0.33 and 3.33% SDD (SDD contained 30% w/w active in its amorphous state). The SDD and excipients (microcrystalline cellulose, anhydrous lactose, croscarmellose sodium and magnesium stearate) were mixed in a low shear blender, followed by co-milling and further blending, and subsequently roller compacted (Alexanderwerk WP120). The resultant ribbons were milled, lubricated and compressed on a rotary tablet machine to a target press weight of 100 mg.

During compression tablets were sampled over five regular intervals and analyzed for content uniformity (total 5 x 10 tablets). Each final blend was riffled and separated into sieve fractions and the active content determined using HPLC for each fraction (>425μm, 180-425 μm, 63-180 μm, <63 μm). A modified Jenike & Johanson segregation test was also employed to evaluate segregation potential. The test used four cycles through a standard hopper design and a 1.8m gravity drop, six 100 mg samples were analyzed for content at different segments of the discharged blend. The physical distribution of SDD in the formulated tablet was also evaluated by FTIR imaging (Thermo Raman DXR). SDD particle size was determined using a Malvern Morphologi G3 and the morphology was confirmed by SEM.

RESULTS

SDD particles produced from the representative spray process were determined to have a mean D50 and D90 of approximately 30-40 μm and 50-60μm, respectively, with a characteristic collapsed sphere morphology. The content uniformity for low dose tablets gave consistently low variation throughout compression , for example the %RSD for 0.1mg tablets remained between 1.5%- 2.4% for all five segments.

The sieve cut potency data, however, showed an uneven distribution of active through the particle size range, with fines (<63um) being super-potent and at least 1.6x nominal content. The data corroborated with the J&J segregation test results which showed an elevated RSD and potency range for the samples analyzed; suggesting a risk for segregation of the final blends on scale up.

Tablet Strength	Mean %RSD Tablet (n=50)	J&J Segregation Results			
		%RSD	Risk*	Potency range (%)	Risk*
0.1 mg	1.97	5.5	Medium	23.6	High
1 mg	1.8	5.5	Medium	21.6	High

**Based on the proposed segregation risk classification system, Shah et. al.*

CONCLUSIONS

Although the theoretical content uniformity model predicted borderline uniformity for the 0.1mg strength product, the experimental data from the manufacture of low dose SDD tablets at a ~10kg scale indicated good dose uniformity is feasible. The content uniformity model assumes that particles are solid, however for hollow or collapsed SDD structures the particles will contain less mass per particle than expected. Therefore, the risk of content uniformity failure will be lower than predicted, with the risk decreasing with lower solid volume fraction (wall thickness). Acceptable uniformity was supported by FTIR imaging showing even SDD distribution on the tablet surface. The granule sieve cut potency and the J&J segregation test results indicate that the risk for segregation remains, and needs to be addressed during process scale-up and bulk transfer operations.

ACKNOWLEDGEMENT

The technical contributions by the Bend Research Inc. team in developing the Spray Dried Dispersions are hereby acknowledged.

REFERENCES

- Rohrs, B.R., Amidon, G.E., Meury, R.H., Secreast, P.J., King, H.M., Skoug, C.J. (2006). Particle size limits to meet USP content uniformity criteria for tablets and capsules. J. Pharm. Sci. 95(5), 1049?1059.

- Shah, K.R et al. (2007)' Assessment of Segregation Potential of Powder Blends', The Pharmaceutical development and technology', *Informa healthcare*, Volume 12, pp 457-462.

(586aa) Consistent Freezing of Bulk Protein Solutions - the Importance of Heat Flow Direction

Rodrigues, M. A., Instituto Superior Tecnico TU Lisbon
Rosa, M., Instituto Superior Tecnico, TU Lisbon
Gomes, D.,
Satish, S. K.,
Geraldes, V., Instituto Superior Tecnico

Purpose

Cryoconcentration during freezing is a major concern for safe storage and transport of bulk protein solutions. As water crystallizes, solutes concentrate at the interface forming a dense layer. This leads to natural convection, causing a generalized heterogeneous distribution of solutes i.e. proteins, sugars, buffer salts. The relative proportions of the solution constituents throughout the ice matrix are dependent of the freezing process and high levels of heterogeneity can compromise protein stability during storage and thaw. These phenomena are greatly dependent of the kinetics and geometry of the freezing process. In this work, we address the influence of the geometry of freezing direction (axial or radial) in the frozen solution's heterogeneity in terms of local concentration of solutes and thermal history.

Methods

Solutions of bovine hemoglobin were frozen radially and axially (from bottom to top) using in small-scale and pilot-scale freezing systems. Concentration of hemoglobin, sucrose and pH were measured trough ice sampling and temperature profiles were measured at several locations. Computational fluid dynamics (CFD), based on the enthalpy-porosity model, was used to predict the ice front velocity, maximum freezable height and concentration of solutes.

Results

The results showed that natural convection is the major cause for the solutes cryoconcentration. A significant improvement in cryoconcentration was observed when the freezing geometry was non-convective, i.e. the freezing front progression was unidirectional from bottom to top. Using this geometry, it was possible to obtain less than 10% variation in solutes concentration throughout the frozen solutions. This result was reproducible, even when the volume was increased by 2 orders of magnitude (from 30 ml to 3 L).The temperature profiles obtained for the non-convective freezing geometry were predicted using a relatively simple CFD model.

Conclusion

The reproducible solutes distribution, predictable temperature profiles and scalability, demonstrates that the bottom to top freezing geometry enables an extended control over the freezing process. This geometry has therefore the potential to contribute to a better understanding and control of the risks inherent to frozen storage, thus providing approaches more consistent with Quality by Design directives.

(586al) Manufacturing Scale Empirical Testing of Modified Packing Procedure for Agarose Based Compressible Protein A Resin

Shrestha, R., Genentech, Inc.
Mendiola, E., Genentech
Cooper, J., Genentech
Smith, M., Genentech, Inc.

Multiproduct mammalian cell culture facilities must maintain operational flexibility, minimizing costly equipment upgrades when possible, while accommodating the evolving needs of downstream purification. High titer monoclonal antibody processes necessitate the implementation of new high capacity Protein A resins as the initial capture step for purification. The column packing requirements for these new resins can present operational challenges. We recently implemented a high capacity agarose based resin with recommended packing conditions to obtain a stable bed that exceeded current equipment capability. Full scale testing (120 cm x 20 cm column) was performed and an alternative packing strategy meeting resin requirements within the existing facility constraints, was identified. The approach taken and the column bed stability assessment performed at scale will be presented.

(606a) High-Throughput Membrane Protein Crystallization Condition Screening Using X-Ray Transparent In Meso Microfluidic Chips

Schieferstein, J. M., University of Illinois at Urbana-Champaign
Khvostichenko, D. S., University of Illinois at Urbana-Champaign
Pawate, A. S., University of Illinois at Urbana-Champaign
Kenis, P. J. A., University of Illinois at Urbana-Champaign
Originally presented on: 11/7/2013 8:30:00 - 8:55:00

The malfunction of membrane proteins has been tied to the initiation and progression of many diseases and, consequentially, membrane proteins account for ~60% of all available drug targets. High-resolution 3-D structures of proteins are essential to understand the structure-function relationship of these molecules and the changes therein that contribute to disease states. However, the amphipathic nature of membrane proteins, the difficulty of expressing and purifying them, and their low yield upon purification have all made them very difficult to successfully screen and crystallize. An emerging technique called the *in meso*method has increased success rates of membrane protein crystallization. This method suffers from limited adoption because of a variety of factors such as the requirement of specialized equipment, expert labor, or expensive robots. Upon successful crystallization, the next bottleneck is the manual handling of the fragile crystals while mounting them on a loop or a capillary, and subsequent cryo-cooling to collect X-ray diffraction data – a process where many crystals suffer damage resulting in poor diffraction data.

To address these challenges we have developed high-throughput X-ray transparent microfluidic chips for *in meso* crystallization screening. The chips are designed with 12-, 24-, and 48- well arrays requiring < 20 nL of protein solution per well and also simultaneously automate the formulation of each sample. X-ray transparency allows these microfluidic chips to circumvent the need for manual crystal harvesting, thus preserving the initial quality of crystals. These *in meso* chips enable the use of lipidic mesophases for crystallization screening on an X-ray transparent platform, a previously unachieved accomplishment. Our microfluidic platforms simplify *in meso* sample preparation and allow for crystallization condition screening, crystal growth and the acquisition of high-resolution diffraction data on chip for structure solution. We have validated our approach on bacterial photosynthetic reaction center and are currently working on challenging membrane proteins from the heme-copper oxidase superfamily.

(606b) Solvent Effect On Crystal Morphology: A Hybrid Approach of a Mechanistic Model and Molecular Dynamics

Kim, S. H., Iowa State University
Doherty, M. F., University of California, Santa Barbara
Originally presented on: 11/7/2013 8:55:00 - 9:20:00

The control of crystal morphology during crystallization processes has a great impact on the end-use properties of the materials (e.g., optical properties) and the downstream performance of the manufacturing process such as the separation characteristics. The crystal morphology depends on external factors such as solvent, supersaturation, temperature, additives and impurities. Various theoretical and computational approaches have been carried out to understand the relation between controlling the crystal morphology and the external growth factors.

Recently, a realistic crystal growth model was developed for non-centrosymmetric solute molecules[1] based on the spiral growth model proposed by Burton, Cabrera and Frank (BCF). This model well explained the non-isotropic behavior of complex drug molecules to predict the crystal shape. In solution growth, the interfacial energies between the solute and the solvent are applied to the spiral growth model. However, the interfacial energies are calculated at the macroscopic scale, so it is hard to relate the microscopic behavior of the interface between the solute and the solvent such as the free energy barriers of the solute molecules that control the rate of incorporation of solute into kink sites.

Molecular simulations have been used to understand the interfacial interactions between solute and solvent molecules at an atomistic level[2] . We applied the atomistic information from the molecular simulations into the mechanistic model of the crystal growth to consider the detailed solvent effect for organic molecules of realistic complexity such as active pharmaceutical ingredients (APIs).

[1] Kuvadia, Z. B.; Doherty, M. F. Spiral growth model for faceted crystals of non-centrosymmetric organic molecules grown from solution. *Crystal Growth & Design* 2011, **11**, 2780

[2] Piana, S.;Gale, J. D. Understanding the barriers to crystal growth: Dynamical simulation of the dissolution and growth of urea from aqueous solution. *Journal of the American Chemical Society* 2005, **127**, 1975

(606c) The Understanding of Nucleation Aspects of Tolbutamide Polymorphs in Solution
Yani, Y., Iowa State University
Chow, P. S., Institute of Chemical and Engineering Sciences, A*STAR (Agency for Science, Technology and Research)
Tan, R., Institute of Chemical and Engineering Sciences, A*STAR (Agency for Science, Technology and Research)
Originally presented on: 11/7/2013 9:20:00 - 9:45:00

The understanding of nucleation aspects of tolbutamide polymorphs in Solution

Yin Yani,*† Pui Shan Chow,† and Reginald B. H. Tan†,‡

† Institute of Chemical & Engineering Sciences, A*STAR (Agency for Science, Technology and Research), 1 Pesek Road, Jurong Island, Singapore 627833

‡ Department of Chemical & Biomolecular Engineering, National University of Singapore, 4 Engineering Drive 4, Singapore 117576

* Corresponding author. Tel: (65) 6796 3852. Email: yin_yani@ices.a-star.edu.sg

Abstract

Experimental works1, 2 have shown that tolbutamide (TB) Form I is preferentially nucleated from solution. Raman spectra collected for TB solute in saturated solution closely matches with that of TB Form I in suspension2. This observation shows that there could be some unique interactions of tolbutamide molecules or solvent molecules in solution that eventually lead to the nucleation of TB Form I. We use molecular dynamics simulation to observe the conformations and arrangements of tolbutamide molecules in saturated and supersaturated solutions. Various solvents were considered in the simulations, such as, ethanol, methanol, and acetonitrile. The simulation results show that, at pre-nucleation stage, TB molecules in solutions arranged close to the structure of TB Form I in vacuum. In addition to that, our simulation has determined that majority of the single TB molecules in solutions arrange more like the type of TB conformation found in Form I.

References

1. Keraliya, R. A.; Soni, T. G.; Thakkar, V. T.; Gandhi, T. R., Effect of Solvent on Crystal Habit and Dissolution behavior of Tolbutamide by Initial Solvent Screening. Dissolution Technologies 2010, 17 (1).

2. Thirunahari, S.; Chow, P. S.; Tan, R. B. H., Quality by Design (QbD)-Based Crystallization Process Development for the Polymorphic Drug Tolbutamide. Crystal Growth & Design 2011, 11 (7), 3027-3038.

(606d) Predicting Protein Solubility and Crystallization Behavior Based On the Second Osmotic Virial Coefficient

Brandenbusch, C., Technische UniversitÃ¤t Dortmund
Herhut, M., Technische UniversitÃ¤t Dortmund
Sadowski, G., Technische UniversitÃ¤t Dortmund
Originally presented on: 11/7/2013 9:45:00 - 10:10:00

The biocatalytic production of complex pharmaceutical proteins (e.g. monoclonal antibodies) has gained an increased industrial interest. Within these processes, the total production costs are often dominated (up to 80 %) by the downstream processing. In state-of-the-art engineering, protein purification is usually achieved by a series of cost-intensive chromatographic steps. One interesting alternative to the conventional processing is protein precipitation, as a first capture step, and crystallization for final product polishing.

In order to develop a precipitation or crystallization step, the solubility of the target protein in aqueous solutions and the type of solid formed (crystal or precipitate) are required as crucial information. In order to supply this data, within this work, an innovative and simple solubility model to measure and predict the protein solubility and crystallization behavior based on the second osmotic virial coefficient (B_{22}) is developed. By use of a potential-of-mean-force model (DLVO) with parameters fitted to B_{22} data at low precipitant concentration, B_{22} at high precipitant concentrations can be efficiently predicted, thus decreasing the experimental effort. By relating B_{22} to the asymmetric activity coefficient, the solid-liquid-phase equilibrium of a protein in solution, consisting of solvent, protein, and precipitant, can be calculated. This combination now enables for the prediction of the protein solubility over a broad concentration range of precipitant with a minimal set of experimental data.

The value of B_{22} herein serves as an ideal measure, as it combines information on the crystallization or precipitation behavior of the protein under the applied conditions, e.g. temperature, pH, type of precipitant used (salt or polymer e.g. PEG) and precipitant concentration.

In order to show the applicability of the model for varying protein sizes, experimental data of B_{22} for two proteins (14 and 144 kDa) in water at varying conditions were measured by static light scattering. The results were compared to experimental and literature data. The results point to the possibility of qualitative and quantitative prediction of the protein solubility in aqueous solution based on two B_{22} and two solubility data at varying precipitant concentrations. This modeling strategy supports a cost-efficient development of crystallization steps.

(606e) Conversion of FBRM® CLD Data to Psd Data: Application to a Needle-Shaped Industrial Case, for the Estimation of Crystallization Kinetics

Mitchell, N., Process Systems Enterprise
Yang, B. S., Boehringer Ingelheim Pharmaceuticals, Inc.
Bermingham, S. K., Process Systems Enterprise
Kim, S., Boehringer Ingelheim Pharmaceuticals, Inc.
Mumtaz, H., Process Systems Enterprise
Originally presented on: 11/7/2013 10:10:00 - 10:35:00

Focused Beam Reflectance Measurement (FBRM®) is a commonly employed Process Analytical Technology (PAT) for tracking the evolution of crystallization processes. FBRM® data can utilised to indicate the onset of nucleation, and the relative rates of other crystallization mechanisms, such as secondary nucleation, crystal growth and/or agglomeration. However, the FBRM® technique does not provide a measure of the in-situ Particle Size Distribution (PSD), but merely the Chord Length Distribution (CLD) of the particles present in the measurement volume of the FBRM® probe. In order to utilise FBRM® data to quantitatively track crystallization processes, a conversion method between FBRM® measurements and PSD data, typically measured via laser diffraction, is necessary. Conversion of CLD to PSD data is not trivial, as numerous process variables, such slurry concentration, crystal shape, refractive indices of the solution and the crystal can significantly affect the measured FBRM®CLD data. This work concerns the application of two conversion methods to an industrial case, which exhibits needle-like crystals. This type of crystal shape represents the most difficult of the application cases for CLD to PSD conversion techniques. Indeed, to the authors' knowledge, all of the published literature on the topic concerns the application of these conversion methods to the more simplistic case of block-like or spherical shaped crystals.

In this work, two conversion methods were applied to the sample industrial case. The first method correlates certain statistics of the measured FBRM® CLD data to an attribute of the corresponding PSD, measured via laser diffraction. This method accounts for the average aspect ratio of the crystals present. In this way, this method accounts for the effect of different crystal shapes on the measured FBRM® data. The second approach utilised in this work, involved the use of the moments (zeroth to sixth), determined from the FBRM® CLD data. These FBRM® based moments were then normalised with respect to the solution concentration and the initial PSD for the seed crystals. This method was previously employed for the simultaneous estimation of the kinetic parameters for secondary nucleation, crystal growth and agglomeration for a paracetamol, methanol and water anti-solvent crystallization process[1]. Both of these techniques were found to provide reasonably accurate (±10% on laser diffraction measurements), indications of certain attributes of the PSD. Finally, the successful application of these conversion techniques allowed the FBRM® data to be quantitatively employed in the estimation of crystallization kinetics for the sample industrial case.

(606f) Problematic Crystal Shapes - Causality and Solvent Selection Guidance Based On Periodic Bond Chain Diagrams

Lovette, M. A., Eli Lilly

Originally presented on: 11/7/2013 10:35:00 - 11:00:00

Crystals with needle like or flat plate like shapes, as typified by aspect ratios ~ 1 : 1 : >100 and 1 : >100 : >100, are often the steady-state growth shapes in the crystallization of pharmaceutical compounds from solution. Crystals with such high aspect ratios may be problematic during subsequent processing steps; e.g., filtration, drying. Furthermore, the crystallization of active pharmaceutical ingredients (APIs) that result in such shapes present additional concerns during the additional drug product processing. Although mechanical techniques such as milling and non-mechanical techniques such as thermal cycling may be able to avoid the isolation of crystals with high aspect ratio shapes, it is often more desirable (especially during early development) to determine appropriate solvents that avoid such shapes altogether. In this presentation, a causality for the formation of needles will be discussed. This causality stems from the three dimensional network of anisotropic intermolecular interactions within a crystal; which are understood using periodic bond chain diagrams. This causality will be applied to different case studies that demonstrate the importance of the dispersive and associative interactions within the crystal and the set of corresponding solvents that are likely abate the formation of needle shaped crystals. From these case studies, a classification system for different types of needles will be provided. The foundations of a similar causality for flat plates and the role of solubility will also be described.

(621a) Engineering Carfilzomib Loaded Nanoparticles for Improved Efficacy and Reduced Systemic Toxicity in the Treatment of Multiple Myeloma

Ashley, J. D., University of Notre Dame
Stefanick, J. F., University of Notre Dame
Alves, N. J., University of Notre Dame
Kiziltepe, T., University of Notre Dame
Bilgicer, B., University of Notre Dame
Originally presented on: 11/7/2013 8:30:00 - 8:52:00

Multiple myeloma (MM), a B-cell malignancy, is characterized by the uncontrolled proliferation of plasma cells in the bone marrow, forming lesions in the bone. While in the bone marrow microenvironment, the MM cells acquire drug resistance by adhering to the bone marrow stroma cells. Current therapies utilized chemotherapeutics to treat MM, however with limited success due to the development of drug resistance. This is a major contributor to the incurability of the disease (median survival 3-5 years), fueling the continued search for better therapies and targets for improved efficacy of treatments.

The proteasome, a class of intracellular proteins responsible for the degradation of proteins, has been recently identified as a viable target for cancer therapy. These proteins also play a vital role in several different signaling pathways within a cell. Malignant cells have been shown to have increased proteasome activity, and thus a greater sensitivity to disruptions in proteasome activity compared to healthy tissue. Carfilzomib, an epoxomicin derivative, is a second generation proteasome inhibitor approved by the FDA for the treatment of multiple myeloma. It irreversibly binds to the chymotrypsin catalytic site of the 20S proteasome, disrupting these pathways and increases the accumulation of pro-apoptotic proteins in tumor cells which results in cell cycle arrest and apoptosis.

Although carfilzomib is an effective treatment, much could be done to improve the therapeutic efficacy. Carfilzomib, like many other chemotherapeutics, suffers from poor water solubility and has to be administered with the aid of sulfobutyl ether beta cyclodextrin (Captisol®) to improve its solubility for clinical use. In addition, studies have shown that carfilzomib causes thrombocytopenia and anemia, among others, in patients. These side-effects are the result of the non-specific interactions of the drug within the body. Limiting these non-specific interactions will reduce these adverse events, improving the therapeutic index and patient quality of life. One means to address both of these issues is through the administration of the drug using a nanoparticle based drug delivery system.

Nanotechnology possesses immense potential for improving the diagnosis and treatment of different cancers. Nanoparticle based drug delivery takes advantage of the enhanced permeability and retention effect created by the "leaky" tumor vasculature. This phenomenon allows for the increased tumor accumulation of nanoparticles having a diameter of 10 - 100nm. The polymer poly(ethylene glycol) is typically used to coat the surface of the particles to provide the particles with the ability to evade clearance by the reticuloendothelial system, allowing for increased circulation time and tumor accumulation *in vivo*. In addition, targeting ligands, such as antibodies, peptides, or small molecules, can be conjugated to the distal ends of the polymer to further enhance the particle's ability to target the tumor cells by binding to receptors on the surface of the malignant cells.

Liposomal nanoparticles are a particularly attractive vehicle for drug delivery. Their facile synthesis and multifunctional capabilities allows them to be used in a wide range of applications. Being generally composed of naturally occurring materials allows them to be biodegradable, a desirable quality in pharmaceutical applications. Therapeutic and diagnostic agents can be loaded in the interior aqueous phase, conjugated to the surface, or embedded in the bilayer of the liposome, allowing for the incorporation of a diverse array of molecules with inherently different properties. This provides a significant benefit especially when this is applied to therapeutics with poor drug-like properties. Nanoparticle incorporation can improve one or more of the drug's properties that hinder its therapeutic effectiveness, such as solubility, bioavailability, or stability, in addition to the advantages of enhanced tumor targeting and circulation properties conferred by the nanoparticle. In addition, incorporating therapeutics within these nanoparticles can reduce the non-specific toxicities associated with the free drug by sequestering the drug within the nanoparticle until its designed release.

In this study, we engineered carfilzomib loaded pegylated liposomal nanoparticles to decrease systemic toxicity while simultaneously enhancing drug efficacy by increasing its accumulation in the tumor. To further increase tumor accumulation and to promote nanoparticle uptake by MM cells, VLA-4 antagonist peptides were conjugated to the engineered nanoparticles to actively target malignant cells via a multifaceted synthetic procedure. The nanoparticles, exhibiting a size of ~70 nm, were internalized by the MM cells and efficiently induced cell death through cytotoxicity. Notably, non-targeting and targeting multifunctional nanoparticles were 2 and 4 fold more efficacious than free carfilzomib *in vitro*, respectively. Most importantly, *in vivo* MM xenograph models established that the nanoparticles demonstrated reduced systemic toxicity with improved tumor growth inhibition compared to the free drug. Combined, this study demonstrates the successful incorporation and administration of carfilzomib loaded nanoparticles with improved therapeutic index for the long term goal of improving MM patient outcomes.

(621b) A Permeation Enhancer for Oral Delivery of Therapeutic Macromolecules

Gupta, V., University of California Santa Barbara
Doshi, N., University of California
Hwang, B. H., University of California Santa Barbara
Mitragotri, S., University of California Santa Barbara
Originally presented on: 11/7/2013 8:52:00 - 9:14:00

Despite its convenience and superior patient compliance, oral route has always been an elusive paradigm for delivery of macromolecules, one of the most sought-after therapeutic interventions for a variety of diseases. Oral peptide delivery is hindered by physiological limitations of the gastrointestinal tract including poor intestinal permeability, and enzymatic/acidic degradation. Although several innovative approaches have been tested, none has reached clinical potential. One such approach includes use of Chemical Permeation Enhancers (CPEs) to modulate intestinal permeability by interacting with epithelial cellular structure. The utility of CPEs however is often hindered by a direct correlation between their potency and toxicity. Here, we report on a novel permeation enhancer, Dimethyl palmitoyl ammonio propanesulfonate (PPS), with excellent enhancement potential and minimal toxicity. PPS has been shown to enhance mannitol transport across Caco-2 epithelial monolayers while being minimally toxic. Continuing the work, PPS was tested for dose- and time-dependent cytotoxicity, delivery of two model fluorescent molecules, sulforhodamine-B and FITC-insulin *in vitro*, and absorption enhancement of salmon calcitonin (sCT) *in vivo*. The studies revealed that PPS enhanced Caco-2 monolayer transport of FITC-Insulin by 2.5-folds in a time- and dose-dependent manner via both paracellular and transcellular pathway. TEER measurements in Caco-2 monolayers confirmed reversible modulation of cellular structure by PPS. Further, PPS enhanced relative bioavailability of sCT by 45-fold following intestinal administration in rats, without inducing irreversible damage to the intestinal epithelium. These results highlight the efficacy and potential of PPS as an excellent permeation enhancer which provides new opportunities for developing efficacious oral delivery systems for therapeutic macromolecules.

(621c) A Systematic Analysis of Peptide Linker Length and Polyethylene Glycol Coating On Cellular Uptake of Peptide-Targeted Liposomes

Stefanick, J. F., University of Notre Dame
Ashley, J. D., University of Notre Dame
Kiziltepe, T., University of Notre Dame
Bilgicer, B., University of Notre Dame
Originally presented on: 11/7/2013 9:14:00 - 9:36:00

Polyethylene glycol (PEG) coated liposomes with defined size range of 100 to 200 nm are widely used vehicles for the selective delivery of chemotherapeutics, most notably for cancer therapy. Liposomal drug formulations benefit from surface-grafted PEG by demonstrating increased stability, bioavailability, and tumor accumulation, while reducing systemic toxicity relative to small molecule drugs. These nanoparticles innately home to tumor sites via the enhanced penetration and retention (EPR) effect through a passive mechanism which can be further improved through active targeting. In an effort to improve tumor targeting and cellular uptake of nanoparticle-based drug carriers, nanoparticles are functionalized with active targeting molecules such as antibodies, antibody fragments, small molecules, and peptides. However, active targeting approaches have not consistently shown successful outcomes, largely due to the wide variety of tumor models employed with variable receptor targets, cellular growth rates, and uptake mechanisms. Additionally, nanoparticle design factors including: traditional methods of generating active targeting nanoparticles, PEG coatings of the particles, linkers used to conjugate the targeting ligands, and the various types of targeting ligands used further contribute to the observed targeting discrepancies.

On liposomal nanoparticles, PEG serves two functions: *i*) to provide stealth to the particles for increased circulation time and *ii*) to act as a linker to connect the targeting ligand to the particle. The current standard for establishing optimal *in vivo* circulation enhancement is the incorporation of a 5 mole percent methoxy-PEG2000-DSPE (PEG2000; a mean of ~45 repeating units of ethylene glycol: EG45) in the liposome formulation. This polymer length and percentage provides complete PEG coating of the liposome surface while maintaining particle stability. In order to create ligand-targeted nanoparticles, targeting ligands have been traditionally grafted onto the distal end of PEG2000. More recently, in an effort to more effectively present ligands above the PEG coating, extended linkers such as PEG3350 and PEG5000 have been successfully implemented. However, long PEG polymers do not preserve a linear conformation in the aqueous phase; instead, they fold within themselves to form globular structures. This unique morphology can bury a large fraction of conjugated ligand and sterically hinder the association of the ligand-targeted nanoparticles with their target receptor. Although PEG itself has been determined to be an ideal molecule to enhance bioavailability, the use of PEG2000 specifically as a liposomal coating and linker has been, in part, due to traditional reasons rather than scientific reasoning. Several reports in literature demonstrate that similar bioavailability profiles and *in vivo* circulation half-life can be achieved with much shorter liposomal PEG molecules such as PEG350, PEG550, and PEG750, despite the estimated PEG coverage of <100%.

Furthermore, a shorter peptide linker may provide a more favorable thermodynamic receptor-ligand interaction by reducing entropic losses associated with long linkers. These provide a strong scientific rationale to evaluate the effect of shorter peptide linker lengths, in coordination with various liposomal PEG coatings, on tumor cell targeting and uptake.

Here, we employed a multifaceted synthetic strategy to prepare peptide-targeted liposomal nanoparticles with high purity, reproducibility, and precisely controlled stoichiometry of functionalities to evaluate the role of liposomal PEG coating, peptide EG-linker length, and peptide valency on cellular uptake in a systematic manner. We analyzed these parameters in two distinct disease models where the liposomes were functionalized with either HER2- or VLA-4-antagonistic peptides to target HER2-overexpressing breast cancer cells or VLA-4-overexpressing myeloma cells, respectively. When targeting peptides were tethered to nanoparticles with an EG45 (~PEG2000) linker in a manner similar to a more traditional formulation, their cellular uptake was not enhanced compared to non-targeted versions regardless of the liposomal PEG coating used. Conversely, reduction of the liposomal PEG to PEG350 and the peptide linker to EG12 dramatically enhanced cellular uptake by ~9-fold and ~100-fold in breast cancer and multiple myeloma cells, respectively. Uptake efficiency is highly dependent on peptide valency, reaching a maximum and a plateau with ~2% peptide density in both disease models. Taken together, these results demonstrate the significance of using the right design elements such as the appropriate peptide EG-linker length in coordination with the appropriate liposomal PEG coating and optimal ligand density in efficient cellular uptake of liposomal nanoparticles.

(621d) Precise Monomeric Control of Multicomponent Random Copolymers: Excipient Design for Oral Drug Delivery

Ting, J. M., University of Minnesota
Navale, T. S., University of Minnesota
Bates, F. S., University of Minnesota
Reineke, T. M., University of Minnesota
Originally presented on: 11/7/2013 9:36:00 - 9:58:00

Random copolymers (RCPs) are macromolecules composed of two or more monomers with desirable functionalities randomly distributed along the polymer backbone. These compounds are instrumental to applications such as the oral delivery of poorly water-soluble drugs, where polymer-drug interplay governs the storage and delivery of active pharmaceutical ingredients. Although binary RCP systems have been studied extensively, there is currently very little guidance on understanding the kinetics of multicomponent copolymers, which are playing an ever-increasing role in imparting more unique physical properties. Herein, we demonstrate a systematic approach for simultaneously targeting exact molecular weights and chemical compositions of multimonomeric RCPs using reversible addition-fragmentation chain transfer free-radical (RAFT) polymerization. An illustrative system to showcase this procedure was studied and compared to a well-known macromolecule utilized in oral drug delivery (HPMCAS, or hydroxypropyl methyl cellulose acetate succinate. The relative reactivities at 70°C between monomer pairs were measured and employed to predict the feed ratio necessary for synthesizing well-defined compositions based on the Walling-Briggs model. Application of the Skeist equations addressed compositional drift and anticipated the general monomer incorporation distribution as a function of conversion, which was verified experimentally. This simple "controlled statistics" paradigm provides complementary synthetic and predictive tools for designing macromolecular chemical architectures with hierarchical control over spatially-dependent structure-property relationships for complex applications such as oral drug delivery.

(621e) Laser-Activated Carbon Nanoparticle Cellular Damage and Prevention

Sengupta, A., Georgia Institute of Technology

Kelly, S. C., Georgia Institute of Technology

Gray, M. D.,

Thadhani, N. N.,

Prausnitz, M. R., Georgia Institute of Technology

Originally presented on: 11/7/2013 9:58:00 - 10:20:00

The cell membrane is a challenging barrier to cross because of its structure and composition. Endocytotic processes can be inefficient, especially for molecules that have therapeutic value such as siRNA, plasmids, etc. So it is a challenging task to come up with a method that can address the issue of intracellular delivery.

Various methods have been employed, like chemical modification of the drug to induce stronger binding to the cell, using vectors to ?infect? cells or using a physical force to temporarily disrupt cells and increase their permeability. The physical methods are sometimes preferred because they are more broadly applicable and involve less modification to the drug. Some examples of these methods use ultrasound, electroporation and lasers for intracellular delivery. The tradeoff associated with using such methods is that generally the forces that cause the cell to permeablize can also cause a fraction of the cells to disrupt, fragment and die. So, there is a need to optimize the process and conditions under which physical methods are operated.

In our previous studies, permeablization of cells was observed when they were exposed to pulsed laser in the presence of carbon nanoparticles.[1] Photoacoustic pressure waves[2], generated as a result of laser-nanoparticle interactions, are believed to cause temporary membrane disruption, which effectively increases permeability. Since the carbon nanoparticles serve as strong laser-absorption sites, the overall absorption capability of the medium increases. This means that much lower laser fluence is required to have an effect on the cells compared to laser delivery systems without nanoparticle absorbers. This kind of drug delivery system is appealing not only because of its novelty, but also because it has the advantages of low exposures and better control of the acoustic field, both of which enable better targeting of drug delivery within the body.

Here we report i) the dependency of uptake and viability on the time of laser exposure, ii) optimized conditions to achieve maximum uptake with minimal loss to cellular viability and iii) potential ways to increase viability of the cells.

A Nd:YAG nanosecond pulsed laser (Continuum Powerlite III Plus, Santa Clara, CA) of 1064 nm wavelength was used for the exposure. The sample holder was made of Pyrex glass (Starna Cuvettes, 37PX-2) since it has high transmittance to infrared wavelength. Human Prostrate cancer cells (DU145) were used as a model cell line in this study. Carbon nanoparticles were made by sonicating carbon black (Black Pearls® 470, Cabot Corporation) in DI water, with Tween 80 was used as a stabilizer, to final carbon concentration of 0.4 g/L. The final suspension was made by adding calcein (10 μM) and carbon black (25 mg/L) to cell suspension containing ~1 million cells/mL in RPMI. Total volume of this solution exposure to laser was 570 μL.

The laser fluence and time were varied for the exposure. After exposure the cells were stored in ice (0 ? 4°C) for at least 5 min until all the exposures were done. The uptake and viability analysis was either done visually using a confocal microscope (Zeiss LSM META/NLO 510) or using flow cytometry (FACS LSR II, BD Bioscience, NJ, USA). To quantify cell fragmentation, we count the number of cell events while keeping the time to flow through the flow cytometer fixed and then compare the number of exposed cells to the unexposed controls. For staining the intact nonviable cells, PI was used. At least three replicates were taken for each experiment. The data acquired as uptake and viability were corrected by comparing them with the control (unexposed cells). Paired t-test was performed to compare between samples. Pluronic solution was prepared by adding 0.5 g Pluronic F-127 to 50 mL DI water, and then thorough mixing at 37°C, followed by sonication for 5 min until a clear solution was formed. The final concentration of the stock was 10 g/L.

When the exposure was set to 18.75 mJ/cm^2 (Figure 1) and time was varied from 0, 1, 3, and 7 min, it was observed that uptake of calcein was observed in 7%, 29%, 61% and 87% of cells, respectively, while viability remained close to 100%. Moreover, when the fluence was increased to 25 mJ/cm^2, the uptake initially went up from 55% of cells at 1 min exposure to 70% at 3 min, and then it fell to 56%. The viability was 87%, 86% and 56% at 1, 3 and 7 min, respectively. At the lower fluence of 18.75 mJ/cm^2, the photo-acoustic pressure field intensity generated as a result of laser carbon interaction should be lower compared to the 25 mJ/cm^2 field. We expect that the cells disrupt less as a result of these waves in the case of the lower fluence. Even when the time of exposure was increased, it did not cause increased loss of cell viability. More cells exhibited uptake of calcein in the presence of the field. In the case of 25 mJ/cm^2, exposure initially at 1 min resulted in about 13% loss of cells with more than half of the cells with calcein in them. Since here the field was much stronger, exposure at longer times resulted in death of cells, which caused the viability to go down, which in turn reduced the uptake of calcein. The optimal uptake was found to be 87%, where DU145 cells with 25 mg/L carbon nanoparticles were exposed for 7 min at 18.75 mJ/cm^2 laser fluence. It should be noted that such efficient uptake with high viability is much better than uptake reported by many other intracellular delivery methods.

Pluronic surfactants have been shown in the past to protect against membrane disruption and damage both in-vivo and in-vitro. Cells were exposed to laser pulses at 44 mJ/cm^2 for 3 min as a negative control. This is a very extreme condition where many cells are killed. Figure 2 shows the effect of adding 5% v/v F-127 Pluronic on the uptake and viability. It can clearly be observed that when no surfactant is added to the system, the viability is about 28% with almost all cells calcein positive, which was expected from such extreme conditions. But when surfactant was added to the system and the same exposure was carried out, the uptake and viability significantly went up to the value of 85% and 90%, respectively. This experiment demonstrates that surfactant is able to save cells from damage.

This study supported the hypothesis that photoacoutic waves generated by laser excitation of carbon nanoparticles cause increased intracellular uptake of molecules into viable cells. The level of uptake and viability depended on laser exposure parameters. At optimal condition in this study, 87% of cells exhibited intracellular uptake of the marker compound, calcein, with cells maintaining almost 100% viability. It was also observed that damage to cells at strong laser exposures can be controlled using pluronics to further optimize uptake with reduced loss of cells. Overall this method holds exciting potential of delivering molecules efficiently into cells with low cytotoxic effect and can be of interest not only for in-vitro but also future in-vivo drug delivery into cells.

REFERENCES

1. P. Chakravarty, W. Qian, M.A. El-Sayed & M.R. Prausnitz, Delivery of molecules into cells using carbon nanoparticles activated by femtosecond laser pulses. *Nature Nanotechnology* **5**, 607?611 (2010)

2. H.X. Chen & G. Diebold, Chemical generation of acoustic waves - A giant photoacoustic effect. *Science* **270**, 963-966 (1995).

ACKNOWLEDGMENTS

This work was carried out at the Center for Drug Design, Development and Delivery and the Institute for Bioengineering and Bioscience at Georgia Institute of Technology with partial financial support from the U.S. National Institutes of Health.

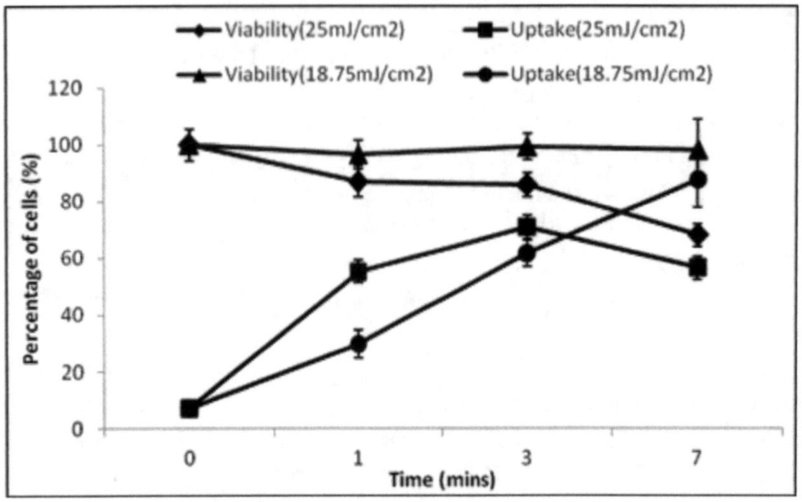

Figure 1: *Photoacoustic effects on intracellular uptake and viability of DU145 cells as a function of laser exposure conditions, using calcein as an uptake marker and carbon nanoparticles at 25 mg/L.*

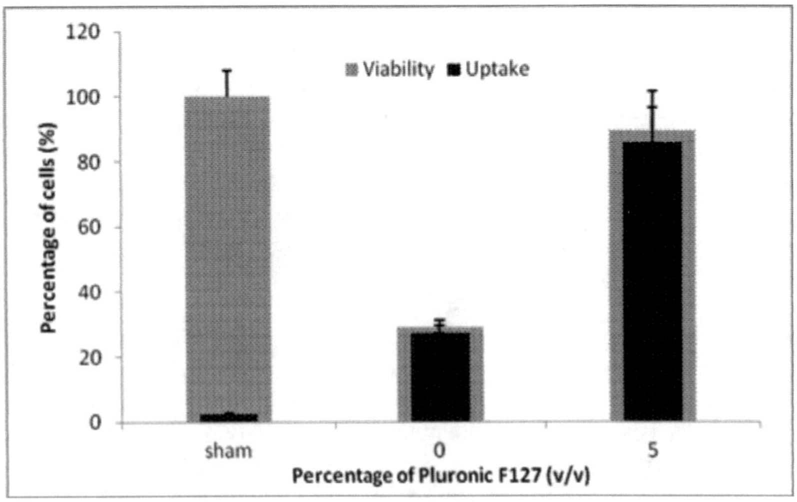

Figure2: *Effect of adding F-127 Pluronic on the uptake and viability of DU145 cells. Cells were exposed at 44 mJ/cm^2 for 3 min.*

(621f) Preparation of Adhesive Mixtures of Surface Modified Glass Beads As Model Carriers in Dry Powder Inhalers and Spray Dried Salbutamol Sulphate

Zellnitz, S.,
Schroettner, H.,
Kappl, M., Max-Planck-Institute for Polymer Research
Urbanetz, N. A., TU Graz
Originally presented on: 11/7/2013 10:20:00 - 10:42:00

In order to use API (active pharmaceutical ingredient) drug particles intended to target the tiny airways of the deep lung in dry powder inhalers (DPIs), they must have an aerodynamic diameter of 1 µm - 5 µm. Particles of this size are rather cohesive and possess poor flow properties [1], which lead to difficulties concerning volumetrically dosing. To overcome this problem, carrier based formulations, where the API is attached to the surface of a larger carrier particle (50 µm – 200 µm) of adequate flowability were developed. Interparticle interactions play a key role in this kind of formulations. On the one hand they have to be high enough that uniform dosing is possible and on the other hand low enough that drug detachment from the carrier surface during inhalation is guaranteed.

As the surface topography of the carrier largely affects interparticle interactions [2], the model carriers used in this work are surface modified glass beads. By physical surface modification in a ball mill glass beads with different shades of roughness have been generated using quartz and tungsten carbide powders [3]. The focus of the present work lies on the preparation of adhesive mixtures of these surface modified glass beads and a model API.

Spray dried salbutamol sulphate was chosen as model API and prepared on a Nano Spray Dryer B-90. It shows the characteristic diameters of x_{10}=0.45 µm, x_{50}=3.07 µm and x_{90}=6.73 µm. Glass Beads in the size range of 400 µm to 600 µm (x_{50}=537.3 µm) processed with quartz (x_{50}=32 µm) for 4 hours and tungsten carbide (x_{50}=25 µm) for 8 hours were used. Adhesive mixtures, with 100% calculated surface coverage of API, were prepared in a tumble blender.

First the homogeneity of the mixtures (expressed as the coefficient of variation of the drug content of n=10 samples out of the mixture and analyzed via high pressure liquid chromatography (HPLC)) was determined and the actual coverage, based on the HPLC results, was calculated. Additionally in vitro lung deposition experiments were performed with a next generation impactor (NGI) and the fine particle fraction (FPF), which is supposed to correlate wit the respirable fraction in vivo, was calculated.

The results show that regardless of what glass beads were used a 100% surface coverage could not be achieved although the amount of API weighed in was calculated for a 100% surface coverage. The actual surface coverage reached for the physically modified glass beads was about 75% compared to a 32% actual surface coverage for the untreated glass beads. By preparing adhesive mixtures with different calculated surface coverage, the calculated coverage and the actual coverage could be correlated for the all the distinct glass beads used and linear equations with different slopes could be set.

Moreover the evaluation of the FPFs shows that the FPF depends on the surface modification, but is independent of the surface coverage.

The obtained data are highly useful to improve the understanding of the mixing behavior of surface modified glass beads, and will help to generate adhesive mixtures whose FPFs and performances can be compared among each other.

Acknowledgements

The authors wish to thank the DFG for financial support within the priority program SPP 1486 "Particles in contact"

References

[1] D.I. Daniher, J. Zhu, Dry powder platform for pulmonary drug delivery, Particuology. 6 (2008) 225–238.

[2] M. Lohrmann, Adhaesionskräfte in interaktiven Mischungen fuer Pulverinhalatoren, PhD thesis University Duesseldorf, 2005.

[3] S. Zellnitz, J.D. Redlinger-Pohn, M. Kappl, H. Schroettner, N.A. Urbanetz, Preparation and characterization of physically modified glass beads used as model carriers in dry powder inhalers., International Journal of Pharmaceutics. 447 (2013) 132–8.

(621g) Effects and Mechanism of Sodium Chloride and Tripolyphosphate On the Formation of Chitosan/Cellulose Sulphate Microcapsules

Zhang, Q., Zhejiang University

Lin, D., Zhejiang University

Yao, S., Zhejiang University

Originally presented on: 11/7/2013 10:42:00 - 11:04:00

Polyelectrolyte complexes (PECs) are association of complexes formed between oppositely charged polymers due to electrostatic interactions. They have gained importance in modern pharmaceutical technology especially for controlled drug delivery. Chitosan is one of the few polycations that are widely used in the pharmaceutical industry due to its high availability, low cost, biocompatibility and easy modification. In order to improve the stability of chitosan PECs, crosslinking agents are introduced. In practice, crosslinking agents may compete with polyanions for binding sites on chitosan. Therefore, the selection of polyanions and crosslinking agents, together with material addition sequences are important factors in PEC preparing procedures. However, little research has focused on the effect of crosslinking agents to the polyanions in the chitosan PEC preparation. In this study, double-layer chitosan PECs were prepared and studied using sodium cellulose sulphate (NaCS) as the corresponding polyanion. Similar systems have been proved for the potential application as oral colon-specific drug delivery systems. Tripolyphosphate (TPP) was used as a crosslinking agent. The effect of TPP on the PEC formation was investigated by changing the TPP addition sequence. Moreover, sodium chloride (NaCl) was used as an adjusting agent in the chitosan/NaCS/TPP PEC formation processes.

The structure of the microcapsules formed under different TPP addition times was characterized and compared. The mechanism of the crosslinking effect of TPP in the chisotan/NaCS system was investigated and the results showed that TPP had the binding priority in the system, which can significantly affect the chitosan/NaCS PECs formation process. Drug release behaviours from these PEC systems were also characterized.

The effects of NaCl on the microcapsule formation process revealed that the microcapsules without NaCl in the formulation had a thin and transparent layer on the surface, which can be easily detached from the solid inner particle that was enclosed inside. However, the microcapsule prepared with NaCl showed a solid particle structure with a thicker wall and good wall-core interaction (Figure 1). The materials constructing both the wall and the core of the two microcapsules were further analyzed using FTIR, SEM, elemental analysis, UV spectroscopy and the mechanism of NaCl effects was discussed. The effect of adding NaCl on controlled drug release was investigated. The results studied can be applied to different PEC systems, which may further facilitate the development of PEC microcapsules for pharmaceutical applications.

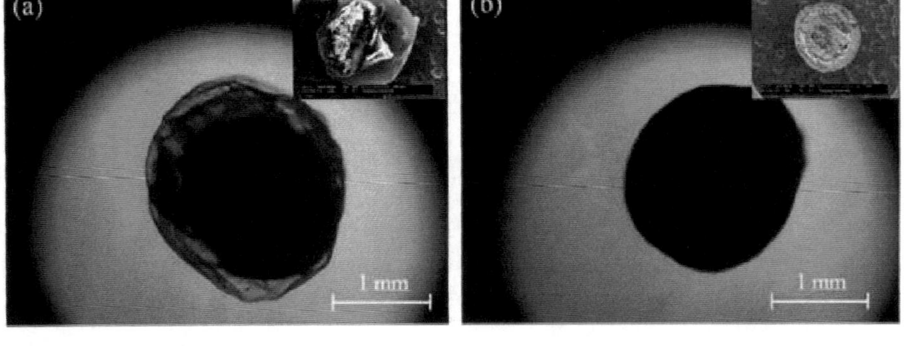

Figure 1. Comparison of microcapsules prepared with or without NaCl. (a) A microcapsule prepared without the addition of NaCl. (b) A microcapsule prepared with the addition of NaCl.

(638a) New Theoretical Methodologies for Pharmaceutical Plant Cleaning

Carr, W., Newcastle University
Martin, E., University of Newcastle upon Tyne.
Talford, M., Britest Ltd
Originally presented on: 11/7/2013 8:30:00 - 8:55:00

Plant cleaning in the pharmaceutical industry is an undervalued but critical stage of processing. Cleaning with solvents and/or surfactants is often the only method capable of removing residual particles. This is often necessary prior to further processing being undertaken in the same equipment. Regulatory bodies have stringent guidelines not only for cleaning equipment but also, as importantly, relating to carryover and/or the potential contamination of subsequent batches of other products from carried over active pharmaceutical ingredients. In a typical multipurpose pharmaceutical plant there are significant implications if cleaning is not achieved effectively and efficiently, ranging from reduced plant availability through to financial penalties and possible plant closure.

Cleaning in pharmaceutical plant is often only considered once the process chemistry has been optimised. The product is regarded as the main source of financial gain in a process, but the question that is rarely asked is - what happens if cleaning is suboptimal? Cleaning challenges can cost companies millions of dollars' as they must clean plant equipment effectively to satisfy regulatory constraints. Failure to do this right first time can result in missed stage gates with processing schedules pushed back, often with their own financial consequences. The empirical approach generally adopted for cleaning protocol development results in plant that is often over cleaned, with cost implications in terms of plant availability, cleaning agent use and disposal.

More reliable methods for plant cleaning are required and thus it is necessary to consider the basic science involved in the production processes carried out prior to defining the cleaning strategy namely (1) structural features of raw materials, intermediates, product or by-products and (2) physiochemical properties of the process materials which may affect cleanability. The research in this presentation describes a new approach to cleaning using multivariate data analysis.

The first approach utilises the fact that cleaning protocols can be determined based on the identification of chemical functional groups within pharmaceutical products. A database of pharmaceutical products was created and a multivariate analysis of the variables related to structure and composition of the functional groups was undertaken. Using principal component analysis (PCA), a set of products were identified which could potentially be cleaned utilising the same approach. This provides a means to determine how products may react to certain cleaning methods with solvents and thus it can be inferred as to how easy or challenging a product may prove to clean from plant equipment.

The second approach is based on the physicochemical properties of process materials. Using a similar PCA-based approach, physiochemical properties are identified that can potentially be used to determine how process materials can be best cleaned from plant and also which products might be difficult to clean from plant equipment.

Both techniques can be used to help determine cleaning protocols based on fundamental scientific principles. Adopting this approach ensures that the cleaning agent used is appropriate to the process chemistry and, therefore cleaning is carried out right first time and as cost effectively as possible. These approaches will enable the inference of worst-case scenario products or intermediates to clean from a processing plant prior to the start of processing. The ultimate goal is for these techniques to be used to support the design of manufacturing processes taking cleaning into account from early stages of development thereby saving time and money during the processing stages.

Beta testing is currently being carried out in collaboration with a number of companies across the pharmaceutical supply chain to validate the practical implementation of this research. Through increased knowledge and understanding of industrial process cleaning, this research will support an industry-wide move towards a target of a more sustainable future.

(638b) Integration of Pervaporation Units in the Production of APIs and Intermediates

Luciani, C., University of Maryland
Buser, J. Y., Eli Lilly and Company
Tharp, G., Eli Lilly & Co
McCulley, M., Eli Lilly & Co
Seibert, K., Eli Lilly & Company
Originally presented on: 11/7/2013 8:55:00 - 9:20:00

Pervaporation is a sophisticated separation method that uses dense polymeric and/or ceramic membranes to isolate one of more components from relatively complex mixtures. This technology offers several advantages over traditional distillation techniques over traditional distillation techniques such as better selectivity, lower energy consumption, azeotrope separation, and modular design. Moreover, pervaporation systems can be considered as green processing practices and they can be integrated to flow configurations. Since it is a diffusion-controlled process, it becomes cost-effective only when the species to be permeated are minor components in the feed. Particularly, in Pharmaceutical Industry, pervaporation systems can be used to remove undesired moisture from reactions matrixes, super-saturate solutions to control crystallizations, dehydrate streams for solvent recovery, remove volatile organic compounds from waste streams, etc.

In this work, two different pervaporation units (PVM-035 and PVM-043, Pervatech) containing ceramic membranes (ceramic-HybSi internal layer) were used to investigate dehydration of organic solvents commonly used in the pharmaceutical industry (e.g., acetone, acetonitrile, methanol, propanol, . Integration of this unit to other unit operations (reactions) was also investigated.

(638c) Injection Moulding As a One-Stop-Production to Produce Pharmaceutical Dosage Forms

Eggenreich, K.,
Schrank, S., Graz University of Technology
Koscher, G., Research Center Pharmaceutical Engineering GmbH
Treffer, D., Graz University of Technology
Roblegg, E., University of Graz
Khinast, J. G., Graz University of Technology
Originally presented on: 11/7/2013 9:20:00 - 9:45:00

Dissolving or dispersing of Active Pharmaceutical Ingredients (APIs) in a molten polymer matrix is a promising method to increase the water solubility of poorly water soluble drugs. The enhanced solubility is achieved by forming various types of solid dispersion. One technology to establish such systems is the Hot Melt Extrusion (HME). Another promising technology which combines the advantages of HME with a direct shaping process is the injection moulding (IM). Thereby the materials are molten by a single screw extruder and fed into an antechamber of the screw. In the next step, the screw stops and acts as a piston applying high pressure on the material and injects it into a mould. The mould consists of different numbers of cavities with the final geometry of the product, e.g. tablets. By defining a specific shape of the cavitie(s), a broad range of geometries can be established. Beside the extrusion, injection moulding is the state-of-the-art technology in the plastic industry. Also in the production of medical devices the IM has a wide field of use.

In our project the focus of investigation is on both the possibilities but also the limitations of IM when used for the production of pharmaceutical dosage forms. The trials were conducted on an E-Mac 50 (Engel, Austria), a small production machine. The experiments followed two different production routes. In the first route, the input materials were pellets made of different polymers (PVCL-PVAc-PEG) and an API in different ratios. They were produced in a prior HME step using a ZSK 18 Twin Screw Extruder (Coperion, Germany). In the second production route, a premix consisting of the API and the polymer was fed directly into the plastification unit of the injection moulding machine. The goal of this route is to establish a one stop production where the raw material is converted into the final product within one production step showing process periods of 2 to 3 minutes.

The obtained tablets were characterized extensively. Product parameters like release behaviour, thermogravimetric behaviour, content uniformity, level of impurities and degradation products, but also physical properties like hardness and friability or the identification of type of solid dispersion were in focus of our investigation.

References:

1. Zema L, et.al.: Injection Molding and its application to drug delivery, J. Control. Release 2012.
2. Quinten T, et.al.: Evaluation of injection moulding as a pharmaceutical technology to produce matrix tablets, European Journal of Pharmaceutics and Biopharmaceutics 71 2009 145–154
3. Breitenbach J.: Melt extrusion: from process to drug delivery technology, European Journal of Pharmaceutics and Biopharmaceutics 2002, 54, 107-117.

(638d) Scaling-Up in Supercritical Fluid Chromatography

Tarafder, A., Waters Corporation
Hudalla, C.,
Iraneta, P., Waters Corporation
Originally presented on: 11/7/2013 9:45:00 - 10:10:00

Supercritical fluid chromatography (SFC) is already recognized as the method of choice for the preparative separation of chiral molecules in pharmaceutical industries. Even for the achiral separations, especially for the separation of natural products, SFC has a huge potential to grow as the method of choice. To harness the full potential of SFC in these separations, however, we need to develop better methods for scaling-up from analytical to preparative conditions.

One of the critical procedures during the method development of a preparative separation is the scale-up procedure. Screening of competing combinations of the mobile phase compositions and the stationary phases, leading to the best separation, is time consuming and expensive to carry out directly in the preparative scale. One of the better alternatives is to first identify the optimum mobile/stationary phase combination in the analytical scale and then transfer the method to a scale where the prep separation should be carried out.

Although there are standard methods available in liquid chromatography (HPLC) for scaling-up, they are not always useful in SFC. In SFC the density of the mobile phase varies significantly across the column, which results into the variation of all the mobile phase properties. The extent and the nature of variation of density in the analytical and the preparative scale can be very different, depending on (a) the dimensions of the connecting tubes and the column, (b) the particle sizes of the stationary phases, and (c) the operating temperature, pressure and flow rate. This difference in the density profiles across the column, between the analytical and the preparative system, will lead to very different chromatography unless the scaling-up procedure is guided by a systematic approach requiring additional calculations. In this presentation we will discuss a systematic approach we are developing to scale-up operations from analytical to preparative scale in SFC.

(638e) Process Transfer and Optimization of a Commercial Roller Compaction Process Leveraging the Maximum Compressive Stress Model and Design of Experiments

Panarello, A., Merck & Co., Inc.
Patel, P., Merck & Co., Inc.
Blair, D., Merck & Co., Inc
Conway, S. L., Merck & Co., Inc.
Harrison, B., Merck Co., & Inc.
Caruso, C., Merck & Co., Inc
Kline, L. A., Merck & Co.
Originally presented on: 11/7/2013 10:10:00 - 10:35:00

In the pharmaceutical industry roller compaction is becoming more attractive as a method of choice for improving material flow behavior and content uniformity over traditional granulation processes. This work herein summarizes the scale-up and tech-transfer of a High Drug Load (HDL) and Low Drug Load (LDL) roller compaction granulation (GRN) to a commercial scale compactor. A Compressive Stresses' model was employed to scale up the LDL & HDL-GRN processes from an Alexanderwerk WP120 to a Fitzpatrick 7x10 compactor. Joint process reviews identified critical gaps and/or commercial challenges and were leveraged to define primary development objectives. During the process validation campaign, poor material flow from the In Bin Hopper (IBH) to the Initial Feed Hopper (IFH) was observed as well as sub-optimized roller compactor operation resulting in various process shut downs and subsequent deviations. A Design of Experiments (DoE) campaign was successfully employed to balance compaction efficiency and achieve the desired GRN-properties.

The Maximum Compressive Stress model was able to define the commercial operating parameters that resulted in GRNs with equivalent material properties (PSD, density, porosity, and compactability) as those manufactured at pilot scale. This enabled the team to rapidly manufacture pivotal Registration Batches at commercial scale. Material balances of the initial commercial scale batches enabled rapid identification of process risks that resulted in optimizing the process to mitigate preferential API holdup within the material transfer system. A correlation can also be made between material losses within the material transfer & compaction unit operations and composite assay. Furthermore, a joint effort was undertaken to successfully execute a commercial scale DoE campaign to optimize the compaction process to reduce equipment stress (i.e., process instability) while yielding GRN properties acceptable for manufacturing bilayer tablets. GRN properties and operating conditions were not influenced by Roll Speed, however; as expected, Roll Pressure did impact granulation properties. Characterization of the Roll Pressure operating space enabled optimization of the process to reduce equipment stress without adversely altering the GRN properties. In conclusion, the LDL and HDL-GRN roller compaction process was successfully qualified and subsequent commercial batches have demonstrated process robustness.

(638f) API Recovery From Pharmaceutical Waste Streams By High Performance Countercurrent Chromatography and Intermittent Countercurrent Extraction

Ignatova, S., Brunel University
Hewitson, P., Brunel University
Sutherland, I. A., Brunel University
Originally presented on: 11/7/2013 10:35:00 - 11:00:00

When new projects are scaled up from the research and development stage into manufacturing, recovery or disposal of waste streams becomes more and more important to minimise cost and reduce both environmental and safety impact. Modern telescopic synthesis techniques and their follow-on purification processes are not always efficient enough to prevent loss of active pharmaceutical ingredients (API) in the waste stream. There are technologies available by which components of such streams can be recovered, treated or disposed of. However, most of them are based on selective filtration and aim to recover catalysts or reduce toxicity of pharmaceutical waste streams before they are released into the general water supply or burnt.

For any waste stream purification, the following stages of process optimization should be considered: prevention and minimising of environmental impact through process design; reuse of materials with no additional processing; recycling materials after such processing if required; disposal with energy recovery and finally benign disposal.

The drive towards greener processes, process economics, technology availability, and legislation, all play a part in the selection of the best available technology for waste stream processing. One such technology is High Performance Counter Current Chromatography (HPCCC) which is being evaluated as new cost effective continuous platform technology as part of a 3 year Technology Strategy Board High Value Manufacturing initiative entitled “Scalable Technology for the Extraction of Pharmaceuticals” [1-2]. This is an industry led collaborative project involving end-users, GSK and Pfizer, a technology supply company, Dynamic Extractions, and Brunel University all actively working together to improve process efficiency and reduce both cost and waste.

HPCCC method development for waste steam purification is done at the analytical scale with subsequent scale up to the preparative 1L instrument. This demonstrates the potential of the technology for the recovery of active pharmaceutical ingredients from complex waste streams in both batch and continuous processing modes. The scalability of the HPCCC process enables optimization work at the laboratory scale to be transferred directly to the 18L pilot scale HPCCC. Different scenarios of throughput/purity/yield will be presented demonstrating the flexibility of the technology to purify the target compound at the required trial grade and avoiding expensive clean up processes. This process is therefore capable of reducing the environmental impact of pharmaceutical waste streams.

References

1. Technology Strategy Board (TSB) High Value Manufacturing Award (Grant No. TP14/HVM/6/I/BD506K) "Scalable Technology for the Extraction of Pharmaceuticals (STEP) September 2009 to August 2012

2. Sutherland, I.A., Thickitt, C., Douillet, N., Freebairn, K., Johns, D., Mountain, C., Wood, P., Edwards, N., Rooke, D., Harris, G., Keay, D., Mathews, B., Brown, R., Garrard, I., Hewitson, P., Ignatova, S. Scalable Technology for the Extraction of Pharmaceutics: Outcomes from a 3 year collaborative industry/academia research programme, *Journal of Chromatography A* **2013**, 1282, 84-94

(681a) Thermal Processing for the Delivery of Water-Insoluble Drugs: Hot-Melt Extrusion and Kinetisol® Dispersing

Williams, R., University of Texas Austin
Lang, B.,
McGinity, J. W., University of Texas at Austin
Miller, D.,
Originally presented on: 11/7/2013 12:30:00 - 12:52:00

Thermal Processing for the Delivery of Water-Insoluble Drugs: Hot-Melt Extrusion and Kinetisol® Dispersing

Robert O. Williams III, Bo Lang, James W. McGinity and Dave A. Miller

INTRODUCTION

With the advances of combinational chemistry and high throughput screening, an increasing number of pharmacologically active compounds have been identified and developed (Lipinski, 2002). A significant proportion of these drug candidates are poorly water-soluble, thereby exhibiting limited absorption. Therefore, advanced pharmaceutical formulation and processing technologies are required to overcome solubility limitations and to enable delivery of these poorly water-soluble drugs.

Amorphous solid dispersion systems are a promising method to improve the wetting and dissolution profile of poorly water-soluble drugs (Serajuddin, 1999). The use of amorphous solid dispersions to enhance the dissolution rate of a drug substance has been employed with several new pharmaceutical products based on oral bioavailability improvement. Formation of amorphous compositions breaks the crystal lattice; therefore, the barrier of drug molecule dissociation is significantly reduced. Moreover, the use of polymeric carriers helps improve the wetting and dissolution properties of hydrophobic drugs with a high partition coefficient (Craig, 2002).

Hot-melt extrusion and Kinetisol® Dispersing are two thermal processing technologies more recently reported for producing amorphous solid dispersions and enhancing the wetting, dissolution and absorption of poorly water-soluble drugs. In the present study, we investigated the two thermal technologies for the production of amorphous solid dispersion systems with drug substances exhibiting different physic-chemical properties.

HOT-MELT EXTRUSION OF ITRACONAZOLE AMORPHOUS SOLID DISPERSIONS

Itraconazole (ITZ) is a broad spectrum systemic antifungal drug exhibiting limited aqueous solubility. Being a weak base, the solubility of ITZ is pH-dependent. At neutral pH values, the de-protonated ITZ showed extremely low solubility. The LogP value of ITZ is 6.5, indicating the highly lipophilic nature of the drug substance. In order to improve the dissolution property of ITZ, amorphous solid dispersion systems have been developed using hot-melt extrusion technology. A series of preformulation studies were performed to select the most suitable polymeric carrier, drug loading, and processing temperature of the hot-melt extrusion process. Drug-polymer miscibility was evaluated using modulated differential scanning calorimetry (mDSC). Results suggested that both Eudragit® L100-55 and HPMCAS LF exhibited good miscibility with ITZ. The glass transition temperatures of binary dispersions were analyzed for drug loading selection. In order to prevent any potential re-crystallization, the glass transition temperature of the amorphous solid dispersion should be at least 50°C above the storage temperature. It was observed that the glass transition temperature could be maintained above 90°C when the drug-to-polymer ratio is 1:2. Therefore, this ratio was kept in the following studies. The precipitation inhibition effect of Eudragit® L100-55 and HPMCAS LF was also evaluated in an *in vitro* assay. Results showed that HPMCAS LF prolonged the period of ITZ supersaturation to a greater extent as compared with Eudragit® L100-55. Therefore, HPMCAS LF was selected as the polymeric carrier. In order to minimize the thermal degradation of drug and polymeric carrier, thermogravimetric analysis (TGA) was performed. HPMCAS LF started to degrade at about 160°C; therefore, the extrusion temperature was controlled at 150°C that was below the onset degradation temperature of the polymer. Hot-melt extrusion was performed on ITZ-HPMCAS system with Vitamin E TPGS as the additive. X-ray powder diffraction (XRD) and DSC studies confirmed the formation of amorphous solid dispersion systems. Dissolution results in fasted state simulated intestinal fluid (FaSSIF) suggested that the ITZ-HPMCAS-Vitamin E TPGS formulation exhibited a significantly improved supersaturation profile as compared with the amorphous solid dispersion without Vitamin E TPGS. As a BSC class II compound, a high degree of drug supersaturation at neutral pH environment helps overcome the dissolution hurdle, which may further enhance the oral absorption of ITZ.

KINETISOL® PROCESSING OF DS901

Kinetisol® Dispersing is a new fusion-based process that was recently developed to rapidly form solid dispersions by imparting high shear and friction forces without external heat input. It enables rapid processing of heat labile drugs and polymers, since these materials are subjected to elevated temperatures for only a few seconds. No solvent or plasticizers are required, although plasticizers have been shown to function as pore formers to increase drug release rates. The absence of a plasticizer has been demonstrated to enhance solid-state stability by producing solid dispersions with higher glass transition temperatures. DS901 is a poorly water-soluble compound with a solubility of less than about 2 μg/mL. Its solubility is pH-independent. The melting point of DS901 is 295°C, raising challenges for the use of hot-melt extrusion. Kinetisol® Dispersing technology was used to produce amorphous solid dispersions of DS901. Amorphous compositions contain DS901, dioctyl sodium sulfosuccinate (DSS) along with four different polymeric carrier systems including HPMCAS LF (a), HPMCAS MF (b), Soluplus® in combination with Eudragit® L100-55 (c), and Eudragit® L100-55 (d). Molecular dispersions of DS901 were obtained. During the Kinetisol® process, all of the compositions were exposed to a temperature of above 100°C for less than 5 seconds. Dissolution results showed that the HPMCAS LF composition exhibited the most promising drug release profile. The lead composition was further tested in an *in vivo* study using Sprague-Dawley rats. The peak blood level of the Kinetisol® composition was 20-times higher than that of the pure DS901 crystalline drug substance.

KINETISOL® PROCESSING OF MELOXICAM

Melocixam is a non-steroidal anti-inflammatory drug with limited aqueous solubility. The solubility in hydrochloric acid solution is 0.9 μg/mL. In water, the solubility of melocixam is 12 μg/mL. The melting point of the drug is 270°C; moreover, meloxicam is thermally labile, causing extreme difficulty for use of hot-melt extrusion processing. As an alternative approach, Kinetisol® Dispersing was investigated for producing amorphous solid dispersion systems of melocixam (Hughey et al., 2011). Polymer screening was performed on povidone, copovidone, and Soluplus®. Results suggested that Soluplus® was the most suitable polymeric carrier for melocixam. Kinetisol® was conducted at a temperature of less than 110°C. An amorphous solid dispersion of meloxicam was formed with minimal degradation. The potency was greater than 97%. As a control group, hot-melt extrusion was also performed at 175°C. Significant thermal degradation of meloxicam was observed and only 88% potency was achieved. Non-sink dissolution testing was performed in different media (i.e., 0.1N HCl and purified water). Results showed that the Kinetisol® composition exhibited a markedly increased dissolution profile as compared with the pure drug substance and a commercial meloxicam tablet formulation.

CONCLUSIONS

As demonstrated in the present studies, thermal processes including hot-melt extrusion and Kinetisol® are feasible for producing amorphous solid dispersion systems to improve the wetting and dissolution properties of poorly water-soluble drugs. Regarding thermally labile drugs and high melting point drugs, Kinetisol® Dispersing is a promising process for rendering amorphous compositions while minimizing thermal degradation of drug substances. The use of hot-melt extrusion and Kinetisol® Dispersing enables the formation of amorphous systems of drugs within a wide window of physic-chemical properties.

Keywords: Hot-melt extrusion; Kinetisol® Dispersing; itraconazole; Meloxicam

REFERENCES

CRAIG, D. Q. 2002. The mechanisms of drug release from solid dispersions in water-soluble polymers. *Int J Pharm,* 231, 131-44.

HUGHEY, J. R., KEEN, J. M., BROUGH, C., SAEGER, S. & MCGINITY, J. W. 2011. Thermal processing of a poorly water-soluble drug substance exhibiting a high melting point: The utility of KinetiSol (R) Dispersing. *International Journal of Pharmaceutics,* 419, 222-230.

LIPINSKI, C. A. 2002. Poor aqueous solubility – an industry wide problem in drug discovery. *American Pharmaceutical Review,* 5, 82-85.

SERAJUDDIN, A. T. 1999. Solid dispersion of poorly water-soluble drugs: early promises, subsequent problems, and recent breakthroughs. *J Pharm Sci,* 88, 1058-66.

(681b) Nano-Extrusion: A Novel One-Step Process for the Manufacturing of Solid Drug Nanoparticle-Formulations

Roblegg, E., University of Graz
Baumgartner, R.,
Khinast, J. G., Graz University of Technology
Originally presented on: 11/7/2013 12:52:00 - 13:14:00

Given the increasing number of poorly soluble active pharmaceutical ingredients (APIs), pharmaceutical scientists increasingly focus on innovative formulation platforms for such molecules. The emerging field of nanoscience, in particular the application of nanosuspensions, offers novel possibilities (1). However, nanosuspensions suffer from stability problems, such as agglomeration. Hence, stabilizers must be used to cover the particle surfaces. Moreover, they are typically delivered parenterally, which is an undesired and invasive delivery route. Thus, it would be beneficial to transform the nanosuspensions into solid oral dosage forms. However, the manufacturing of solid-nanoparticle formulations requires several challenging steps. Five products (Rapamune®, Emend®, TriCor®, Megace®ES and Triglide™) produced via nanoparticle technology are currently available on the market, of which four are based on the media milling technique (Elan NanoCrystal® technology). The remaining one (Triglide™) is manufactured using high-pressure homogenization (DissoCubes®, SkyePharma PLC) and requires several steps, including freeze drying, spray drying, spray coating and pelletizing or granulation, to convert the nano-particles in a solid powder for compaction into tablets or filling into capsules (2). Since these complex processes are time-consuming and cost-intensive, new technologies are needed to transfer a nano-suspension into a solid application form while avoiding any agglomeration. In the present study, a one-step nano-extrusion process is presented where the nanosuspension is directly fed to a hot-melt extruder. The goal was to obtain extrudates that contain homogenously distributed and de-aggregated embedded API nano-crystals in a polymer matrix (3). Using this novel technique, the drug release rate of the poor soluble drug was significantly enhanced.

Phenytoin, a poorly soluble antiepileptic drug with a melting temperature above 280°C was used as model drug. Stable aqueous nanosuspensions of Phenytoin were produced via high-pressure homogenization by adding appropriate stabilizers (i.e., Tween® 20 and Kolliphor® EL) in accordance to the physicochemical characteristics of the API. Next, the nanosuspension was fed to the extruder via a liquid side feeding device, while using Soluplus® as matrix-material (glass transition temperature 70°C). The solvent was removed continuously via devolatilization, which is standard procedure during extrusion of solvent-containing drug-excipient mixtures. The obtained extrudates were investigated via TEM and EDX to visualize and verify the embedded nanocrystals. Furthermore, dissolution studies were conducted to evaluate the drug release profile.

The addition of both stabilizers to Phenytoin-dispersions results in stable nanosuspensions, which were added into the molten polymer via side feeding of a hot melt extruder. TEM- and EDX data of the obtained extrudates confirmed that the nanocrystals were embedded in the matrix in a de-aggregated form.

Furthermore, the dissolution studies displayed an improvement of Phenytoin release from the solid nanoparticle-formulation in comparison to the untreated Phenytoin-powder.

The current study demonstrates that this novel process based on hot-melt extrusion of a nanosuspension is an appropriate continuous technology for producing solid nano-formulations. This one-step process helps to eliminate problems associated with stabilization of nanosuspensions and the conversion of nanoparticles into a solid dosage form. Furthermore, the embedded drug nanocrystals allow an improvement of the dissolution properties of poorly soluble and therefore poorly bioavailable APIs.

1. Müller RH, Jacobs C, Kayser O. Nanosuspensions as particulate drug formulations in therapy. Rationale for development and what we can expect for the future. Adv Drug Deliv Rev. 2001; 47(1): 3–19.

2. Van Eerdenbrugh B, Van den Mooter G, Augustijns P. Top-down production of drug nanocrystals: nanosuspension stabilization, miniaturization and transformation into solid products. Int J Pharm. 2008; 364(1): 64–75.

3. Khinast J, Baumgartner R, Roblegg E. Nano-extrusion: a One-Step Process for Manufacturing of Solid Nanoparticle Formulations Directly from the Liquid Phase. AAPS PharmSciTech. 2013; published online: March 06

(681c) Can You Believe Your Eyes? High Comfort Contact Lens Wear Around the Clock

White, C. J., Auburn University
Bartel, C. J., Biomimetic & Biohybrid Materials, Biomedical Devices, & Drug Delivery Laboratories, Department of Chemical Engineering
Thomas, C. R., Biomimetic & Biohybrid Materials, Biomedical Devices, & Drug Delivery Laboratories, Auburn University
Byrne, M. E., Auburn University
Originally presented on: 11/7/2013 13:14:00 - 13:36:00

Comfort agents are commonly used in eye drops to sequester and retain water within the ocular tear fluid, while enhancing other comfort contributing properties, such as tear viscosity, tear film stability, and lubrication. In addition, incorporating comfort agents into various ophthalmic devices, such as corrective contact lenses, significantly increases ocular comfort and biotolerance. A novel contact lens material was engineered for the controlled, simultaneous release of trehalose, a low molecular weight disaccharide, and 120 KDa hydroxypropyl methylcellulose (HPMC). Through molecular imprinting, a high degree of control over the release of both trehalose and HPMC was achieved, the first demonstration of simultaneous release of comfort agents from the same lens. Trehalose has been shown to provide high levels of comfort to the eye when delivered from eye drops, but this is the first demonstration of release from contact lenses. Release of trehalose was controlled to provide high levels of comfort during the first half of lens wear by sequestering high levels of water within the tear fluid to prevent dry eye discomfort. End of day discomfort is provided by the extended release of HPMC, which maintains lens lubrication, refreshes the lens surface, and prevents lens dehydration as well as lowers protein adhesion to the surface of the lens. By applying the principles of molecular imprinting to a silicone hydrogel contact lens, high level of control over the release of trehalose and HPMC was achieved. Increasing the molar ratio of acrylic acid to comfort agent (M/T) demonstrated independent control over the release of both HPMC and trehalose, extending the trehalose duration of release by approximately 5,000%. In addition, use of molecular imprinting improved loading, optical clarity, and physical properties of the lens. For the first time, incorporation of HPMC was shown to lower protein adhesion to the surface of the lens by up to 30%. By utilizing trehalose to provide high initial wear comfort and HPMC to ensure end of day lens comfort, it is hypothesized that the maximum level of lens wear comfort can be achieved. To demonstrate the high level of comfort provided by this lens, a novel rating system was developed based on previously undetermined relationships between comfort properties and the effect of different comfort agent species, comfort agent solution concetration, and comfort agent molecular weight.

(681d) Development of a Coating Process to Apply Water-Insoluble Drugs On Microneedles Using Molten Coating Solutions

Ma, Y., Texas Tech University
Gill, H., Texas Tech University
Originally presented on: 11/7/2013 13:36:00 - 13:58:00

Microneedles is a promising approach to deliver potent drugs via the skin and other easily accessible parts of the human body. The coated microneedle approach for drug delivery utilizes the microneedle as a carrier of the drug in the form of a thin film coated on the microneedle surface. Insertion of this drug-coated microneedle in the tissue enables drug delivery via dissolution of the water soluble film in to the interstitial fluid. This approach relies on the use of aqueous solutions with water-soluble drugs and coating excipients. To expand the coated microneedle approach to include delivery of lipophilic drugs, we present for the first time the process of coating water-insoluble drugs using molten coating solutions. Polyethylene glycol (PEG) was chosen as a drug solubilizer and emulsifier to form a matrix system, which allows the water-insoluble drug to be dispersed in the molten PEG phase through solubilization or emulsification. Lidocaine, a commonly used local anesthetic was chosen as a model drug to test the feasibility and consistency of the molten coating process. A custom designed microprecision automated coating system was used to fully control the temperature of the molten coating solution, microneedle dipping depth, and coating speed. The results of our study show that the coating films on microneedles are uniform and could provide substantial amount of drug on microneedles. In addition, the PEG matrix could significantly enhance dissolution of the water-insoluble lidocaine in to the tissue. In conclusion, the molten coating process we introduce here offers a practical approach to coat water-insoluble drugs on microneedles.

(681e) Nanoparticle-Stabilized Colloids in Compressible Hydrofluoroalkanes

Wu, L., Wayne State University
Conti, D. S., Wayne State University
da Rocha, S. R. P., Wayne State University
Originally presented on: 11/7/2013 13:58:00 - 14:20:00

Hydrofluoroalkanes (HFAs) are the propellants of choice in a broad range of industrial applications, including medical sprays such as pressurized metered-dose inhalers (pMDIs). pMDIs are of relevance in the context of oral inhalation (OI) therapies as they are the least expensive and most widely used devices for the regional delivery of drugs to the lungs. The potential of pMDIs is much broader, however, as the lungs are also recognized as an efficient pathway for both small molecular weight drugs and therapeutic biomacromolecules to the blood stream. However, most therapeutics have limited solubility in HFAs, due to their dual hydrophobic and at the same time lipophobic character, typically requiring the formulation of drugs as colloidal dispersions, or the use of co-solvents to enhance drug solubility for the formulation of solution pMDIs.

In this work we show that nanoparticles (NPs) dispersed in compressible hydrofluoroalkanes (HFAs) at small volume fractions are capable of stabilizing micron-sized particle colloids, which otherwise flocculate due to strong van der Waals forces. Water-soluble, biodegradable NPs with a chitosan (CS) core, grafted with highly HFA-philic moieties can be readily dispersed in the low dielectric HFAs, and are capable of imparting stability to a wide range of therapeutic particles having different chemistries (polar or hydrophobic; small and large molecular weight, including peptides and proteins) and morphologies (micronized crystals or amorphous), and producing aerosols conducive to deep lung deposition. These NP systems thus serve as a broadly applicable platform for the non-invasive delivery of therapeutics to and through the lungs using propellant-based, portable inhalers, and are also of potential relevance in other industries where HFAs are employed as solvents or propellants. This concept may also be applicable to other compressible solvents.

(681f) Applications of Affinisol HPMC in Hot Melt Extrusion

O'Donnell, K.,
Porter, W. III, The Dow Chemical Company
Shrestha, U.,
Reinhardt, A.,

Originally presented on: 11/7/2013 14:20:00 - 14:42:00

The objective of this study was to assess the ability of AFFINISOL™ HPMC to be processed by hot melt extrusion and to determine the influence of molecular weight grade on dissolution rate and supersaturation maintenance. AFFINISOL™ HPMC and griseofulvin were pre-blended in a Turbula mixer for 10 minutes and extruded via a Leistritz Nano16 hot melt extruder utilizing a 2 mm strand die. Resulting strands were either milled in a Fitz mill at 9000 rpm or pelletized via a ConAir pelletizer. Milled extrudates were physically characterized by Differential Scanning Calorimetry (DSC), Thermogravimetric analysis, and powder X-Ray Diffraction (XRD). Dissolution was performed in pH 5.8 phosphate buffer for 24 hours in a USP II dissolution apparatus with a paddle speed of 50 RPM. Dissolved drug was quantified by HPLC. Thermogravimetric analysis revealed no early onset of degradation at the temperatures used during extrusion. All formulations were successfully extruded below the melting point of the API into 2 mm strands which were translucent in nature with minimal color change observed. All extrudates were amorphous as evidenced by the absence of any crystalline peaks in the XRD diffraction patterns or melting endotherms in DSC thermograms. All molecular weight grades achieved and maintained supersaturation over the 24 hour testing period. Dissolution rate was dependent upon molecular weight grade utilized with higher molecular weights resulting in slower dissolution. AFFINISOL™ HPMC can be successfully processed by hot melt extrusion and is capable of generating amorphous solid solutions. Release rate of the API can be controlled by varying the molecular weight. By using hot melt extrusion and selecting the appropriate molecular weight it is possible to get both controlled release and supersaturation concurrently.

(695a) Life Post Approval: A Real Life Example

Seymour, C., Pfizer Inc

Originally presented on: 11/7/2013 12:30:00 - 12:55:00

Quality by Design poses challenges and opportunities post NDA and MAA approval. This presentation will explore post approval Quality by Design challenges and opportunities. Case studies will be presented that highlight the recent learnings and challenges in the post approval area from the technical perspective as well as the regulatory perspective.

(695b) The Application of Chemical Engineering Tools to Post-Launch Process Optimization

Glennon, B., UCD
Carr, M., APC Ltd
Barrett, M., University College Dublin
Originally presented on: 11/7/2013 12:55:00 - 13:20:00

As pressure grows on companies to deliver increasing levels of process performance and improvement, there is an increased role to play for more effective application of chemical engineering principles to process design and optimization.

In this talk, we will present several case studies discussing the use of advanced process modelling, measurement and control to deliver significant process improvements to existing processes, often within the confines of the regulatory filing. Particular emphasis will be placed on the delivery of improved particle attributes through better reaction engineering, unit operation modelling, crystallization design and isolation optimization.

(695c) Scale-Up and Operation of Continuous Hydrogenation in the Manufacturing Setting

Spencer, R., Eli Lilly and Company

Originally presented on: 11/7/2013 13:20:00 - 13:45:00

Eli Lilly has been developing Continuous technologies for drug substance manufacure for several years. This technology is now being scaled up and transferred from the development setting to the manufacturing setting. The technical and operational considerations in scaling up a Continuous Hydrogenation in a manufacturing setting will be discussed. This includes operational, safety and robustness items aswell as technical scale-up issues.

(695d) Monitoring Batch-to-Batch Reproducibility of TCM Production Process Using in-Line NIR Spectroscopy Combined With MSPC Method

Li, W.,
Qu, H., Zhejiang University
Originally presented on: 11/7/2013 13:45:00 - 14:10:00

Monitoring batch-to-batch reproducibility of TCM production process using in-line NIR spectroscopy combined with MSPC method

Wenlong Li, Haifan Han, Haibin Qu *

(*Pharmaceutical Informatics Institute, Zhejiang University, Hangzhou 310058, China*)

Traditional Chinese medicine (TCM) products are usually manufactured through batch processes. To improve the batch-to-batch reproducibility, the feasible approaches for real-time monitoring of batch evolution need to be developed. In-line near-infrared (NIR) spectroscopy combined with multivariate statistical process control (MSPC) method as an efficient process analytical technology (PAT) tool, is presented in this study for the real-time batch process monitoring. Two representative technical processes, the alkaline precipitation of the compound E Jiao oral liquid and the liquid preparation process of Tanreqing injection, were taken as examples. The process NIR spectral data (three-way matrix) was unfolded in "variable-wise" way into the two-dimensional matrix, and multi-way principal component analysis (MPCA) model were developed based on the rearranged data of the normal operation condition (NOC) batches. Three kinds of multivariate control charts (scores, Hotelling $T^2$and DModX) were used to monitor the evolution of test batches with artificial batch variations, including the change of starting material quality attributes and abnormal operation conditions.

Case 1: Alkaline precipitation is a critical unit operation used in the manufacture of compound E-Jiao oral liquid, which has the recognized curative effect of invigorating vital energy and nourishing blood. The in-line NIR spectra of 29 batches of feed liquid, including 13 NOC batches and 16 AOC (abnormal operation conditions, AOC) batches, were collected using an Antaris II Fourier transformation (FT) NIR spectrometer (Thermo Electron Co., Madison, USA) with a handheld fiber-optics probe over the wave number range of 4000- 10,000 cm^{-1} at 4 cm^{-1} data interval. The MSPC models were developed using the SIMCA 13 software (Umetrics, MKS Instruments Inc., Sweden), and the effectiveness of the proposed approach was evaluated through experimental verifications. 10 batches run under normal operating conditions were used to study the batch-to-batch variation and to definite the MSPC control limits. Another 3 NOC batches and 16 AOC batches, which were operated with abnormal densities of the raw materials, or under abnormal pH values, solid-to-liquid ratios, and heating-up temperatures, were used as the test batches to challenge the established models. As illustrated with test batches, the established monitoring model can identify NOC or AOC batches accurately, and detect kinds of deviations from NOC batches using the control charts. The consistent monitoring result was derived according to the score and Hotelling T^2control charts, and the two types of control charts could be individually or collectively used for monitoring batch evolution.

Case 2: The liquid preparation process of bear bile powder extract is a critical operating unit during the manufacture of Tanreqing injection, which is used chiefly in treating infection of the upper respiratory tract and serious influenza. The simulation experiments were designed, including 8 NOC batches and 7 AOC batches (with abnormal solid-to-liquid ratios, sodium hydroxide dosage, heating-up temperature, or stir speeds). Spectra were in-line collected from a simulation device using an Antaris MX Fourier FT-NIR spectrometer (Thermo Electron Co., Madison, USA). The MSPC model for the in-line monitoring of liquid preparation process of bear bile powder extract was developed after the selection of 6 NOC batches as calibration set. The fault monitoring was performed through the sore charts, Hotelling T^2charts and DModX charts, with the average values of the calibration batches as "golden batch trajectory" and ±3 standard deviations as control ranges. The trajectories of the test NOC batches were all lie in the control range, and the trajectories of the test AOC batches were all out of the controlled range in various ways. The results suggested that the established method can be used for the in-line monitoring and control of the fluid preparation process of bear bile powder extract effectively.

The batch process monitoring approach presented in this study, using in-line NIR spectroscopy combined with multivariate data analysis, was found very effective for the real-time monitoring of process deviations from NOC batches. It is an alternative promising tool for monitoring batch reproducibility of the unit operations during the manufacture of TCM, which can enhance process understanding and open up the possibility of process control to achieve the desired product quality in the manufacture of TCM. The key to establish a satisfactory MSPC model is to select good batches representative of normal operating conditions for model training and to establish appropriate control limits. To further diagnose the causes of process abnormalities, contribution plots could be used in combination with the control charts to identify NIR wavelengths associated with a process fault/raw material change, and additional models are also needed to be established between the process parameters and NIR spectral data.

*Corresponding author. Tel.: +86 571 88208428; fax: +86 571 88208428.

E-mail address: quhb@zju.edu.cn (H. Qu).

(695e) In-Line Monitoring of the Extraction Process of Panax Notoginseng (Burk.) F. H. Chen Based On the Quantitative Use of Near-Infrared Spectroscopy

Zhou, C., Pharmaceutical Informatics Institute, Zhejiang University
Li, W., Pharmaceutical Informatics Institute of Zhejiang University
Qu, H., Zhejiang University
Originally presented on: 11/7/2013 14:10:00 - 14:35:00

Panax notoginseng (Burk.) F. H. Chen, a well-known medicinal herb that belongs to the ginseng family, has a long history of use in China. It has been traditionally used for its hemostatic and cardiovascular effects, which help arrest internal and external bleeding, reduce swelling and pain, as well as to disperse blood clots, eliminate blood stasis and promote blood circulation. The pharmacological properties of *Panax notoginseng* are generally attributed to its saponins, including ginsenosides, notoginsenosides and gypenosides. Among these saponins, arasaponin R_1 and ginsenoside Rg_1, Re, Rb_1, Rd are considered to be the major active compounds.

The extraction process of *Panax notoginseng* was the first step of *XueShuantong* injection manufacturing. It was considered to be a crucial process as it defined the quality and quantity of the effective constituents in the final product.

Monitoring extraction solutions with NIR spectroscopy will produce an absorption spectrum of the chemicals present in the extraction solution in a rapid and non-destructive way. These spectra can then be manipulated using multivariate data analysis techniques to develop calibration models for each measured variable. Although the initial model building will require reference data based on the traditional destructive analytical methods, a robust model can thereafter be used to predict the quality attributes in a non-destructive way.

For NIR measurements, the fiber optic probe was immersed in the extracting solution to collect NIR spectra in real-time. Ultra performance liquid chromatography (UPLC) was used as a reference method to determine the concentrations of arasaponin R_1 and ginsenoside Rg_1, Re, Rb_1, Rd in the extraction solution. Partial least squares regression (PLSR) calibration models of arasaponin R_1 and ginsenoside Rg_1, Re, Rb_1, Rd were built separately, using different spectral regions and pretreatments. The results showed that the correlation coefficients of inner cross validation were 0.9962, 0.9902, 0.9915, 0.9951 and 0.9876, and the root mean square error of cross validation (RMSECV) were 0.057, 0.232, 0.028, 0.106 and 0.084 (mg/ml), respectively. The established models were used for in-line and real-time monitoring of a new extraction batch, and the root mean square error of prediction (RMSEP) were 0.0568, 0.233, 0.0411, 0.168 and 0.0621 (mg/ml), respectively. The NIR spectroscopy was proved to be an effective technology in the real-time monitoring of extraction process of *Panax notoginseng*, providing real-time information of the extraction process.

(758a) Development and Implementation of a Process to Control Off-Gassing

Lee, A. G., Bristol-Myers Squibb
Lo, E. T., Bristol-Myers Squibb Company
Natalie, K., Bristol-Myers Squibb Co,
Vu, T., Bristol-Myers Squibb Company
Rossano, L., Bristol-Myers Squibb Company
Originally presented on: 11/7/2013 15:15:00 - 15:35:00

The use of the *tert*-butyloxycarbonyl (BOC) moiety is a highly effective protection strategy for amines in the synthesis of small molecule active pharmaceutical ingredients. While evaluating these chemistries, the evolution rate of carbon dioxide and other volatile byproducts from these reactions is a major consideration for scale up. In our development of a BOC protection reaction to produce a pharmaceutical intermediate, the physical property of a solid intermediate severely hindered mixing in the vessel. As the reaction progressed, changes to the solid's physical property resulted in rapid uncontrolled off-gassing. We discuss the development of a process using acid and base charge equivalents, charge rate and pH to address the solid intermediate formation and gas evolution. We scaled up the process in the kilo plant to demonstrate a range of process parameters which yielded good control over gas evolution while maintaining acceptable quality and yield.

(758b) Chemical Engineering Strategies to Directly Deliver Process Properties Across Development and Manufacturing Scales

Glennon, B., UCD
Carr, M., APC Ltd
Whelan, J., University College Dublin
Barrett, M., University College Dublin
Originally presented on: 11/7/2013 15:35:00 - 15:55:00

In recent years there has been a growing necessity to reduce the time, cost and risk associated to the development and scale-up of pharmaceutical processes. The increasing pressures to direct deliver critical quality attributes (CQAs) and process performance criterion (cycle times, capacities etc) across manufacturing and supply scales compounding this challenge. As a result of this process engineering strategies, to enable streamlined, reliable and robust scale-up, has become increasing important.

Within this presentation novel process engineering strategies and technologies, underpinned by process modelling, will be presented that enable the delivery of process characteristics (i.e. particle size, shapes, PSDs, flowability, reaction purities, process kinetics etc). Applications of advanced process control, process simulation, kinetic assessments, particle engineering strategies, processing technologies (batch and continuous) will be discussed via numerous development and scale-up case studies.

(758c) The Chemical Engineer Strikes Back â€“ A New Hope for Commercial Process Development
Rothstein, S.,
Originally presented on: 11/7/2013 15:55:00 - 16:15:00

The process challenges of optimizing and scaling up the reaction to make 2,6-diaminopyrazine-1-oxide (DAPO), a precursor to a highly promising insensitive energetic material, will be discussed. The original process featured a precisely controlled dual addition of a liquid base and a solid reagent to limit the formation of more than 30% impurity and minimize starting material decomposition. These conditions limited this labor intensive reaction to the laboratory scale and produced DAPO with inconsistent yield and purity. Our understanding of the chemical reactivity led to improved reaction conditions featuring liquid doses rather than a solid dose, thus allowing for the rate of the reaction to be controlled by automated dosing. Using a kinetic model built with DynoChem® software, a robust process was optimized to improve the yield of DAPO and significantly reduce impurity formation. The reaction was scaled up using the optimized conditions and provides a potential commercial process for DAPO.

(758d) Reaction Engineering for a Sulfonamide Intermediate

Marek, J. C., AbbVie, Inc.
Zhao, Y., AbbVie, Inc.
Hill, D. R., AbbVie, Inc.

Originally presented on: 11/7/2013 16:15:00 - 16:35:00

Reaction engineering of a sulfonamide synthesis from early through mid and late stage development is discussed, with emphasis on consideration of reaction pathways. Reaction of sulfonyl chloride with ammonia in acetone solvent eliminates the use of aqueous ammonium hydroxide at extreme cryogenic temperatures (-60 °C), or for solid ammonium carbamate to achieve the amidation. The order of addition in the semi-batch amidation reaction was chosen to minimize use of excess equivalents of amidation reagent. In the revised process, the ammonia / acetone solution is charged into a solution of sulfonyl chloride, maintained at sub-ambient temperature (NMT -10 °C). Hence, both temperature and reagent equivalents are used to control aniline sulfonamide impurity levels. Formerly, the sulfonyl chloride was added to an excess of ammonium carbamate reagent at sub-ambient temperatures, and was promptly quenched to avoid over-reaction. The revised process affords robustness in design space including less potential for over reaction to aniline impurity.

(758e) Scale-Up Oscillatory Helical-Baffled Reactor for Multiphase Reactions

Phan, A. N., Newcastle University
Harvey, A. P., Newcastle University
Originally presented on: 11/7/2013 16:35:00 - 16:55:00

Scale-up oscillatory helical-baffled reactor for multiphase reactions

Anh N. Phan and Adam P. Harvey

School of Chemical Engineering & Advanced Materials (CEAM), Newcastle University, NE1 7RU, UK

Corresponding to anh.phan@ncl.ac.uk

Mesoscale oscillatory helical-baffled reactor (MOHB), a helical wire embedded inside a smooth tube without modification to the tube, was constructed and characterised at very low net flow rates (net flow Reynolds numbers, $Re_n \leq 10$) over a wide range of oscillation conditions. The fluid mixing is controlled by the oscillatory flow and is thus independent of net flow. The flow becomes chaotic at high oscillatory Reynolds numbers and the secondary flow (swirling) becomes active, which imposes an axial velocity component. The coupled effect of the vortex flow and the swirling flow in the presence of helical baffles augments the significant radial mixing and greatly broadens the operating/design window of these reactors. Plug flow performance can be achieved at a ratio of oscillatory Reynolds number to net flow Reynolds number of approximately 10-250 and a helical turn to tube diameter ratio in a range of 0.2-0.6. The highest degree of plug flow was achieved at Strouhal number $0.2 < St \leq 0.1$ and Re_o=400-700 with the number of tanks in series equal to the number of helical turns. At a scale of 5mm inner diameter, the MOHB is considered to be effective and robust in scale-down production of high added value products, decreasing the reagents requirements and waste.

Is there a correlation between this mesoscale and a conventional scale? We scaled the mesoscale reactor up to 13 times and 130 times in volume by maintaining geometrical parameters (opened cross-sectional area and helical turn to tube diameter ratio) and oscillatory conditions.Â It was found that the same mixing behaviour in terms of residence time distribution profiles of tracers was obtained in the large scales as in the mesoscale.

Keywords: scale-up, helical-baffled reactor, oscillatory flow, plug flow

(758f) Process Scale-Up From Drying of a Single Drop to Production Scale Spray Dryer Using Levitated Single Droplet Analysis (DKA) and Computation Fluid Dynamics (CFD)

Walter, M., GEA Process Engineering Inc.

Originally presented on: 11/7/2013 16:55:00 - 17:15:00

Process Scale-Up from Drying of a Single Drop to Production Scale Spray Dryer Using Levitated Single Droplet Analysis (DKA) and Computation Fluid Dynamics (CFD)

Since its recent introduction, the single droplet analyzer (DKA - Digital Kinetics Analyzer) and Computational Fluid Dynamics (CFD) has allowed optimization of spray dryer design.Â Prior to utilizing DKA combined with CFD, one would have to theorize how a powder would behave during drying under various conditions and design the spray dryer accordingly.Â Â New Computational Fluid Dynamics modeling power combined with a novel means to examine in detail how a single drop of feed dries has allowed spray dryers to be more efficient, safer, smaller, of higher capacity, and with fewer operational headaches than in the past.

In addition to improvements in spray dryer design, DKA allows us to examine how individual particles dry and how different drying parameters affect the dried product.Â This allows the spray dryer to produce a ? custom particle? through adjustment of spray drying parameters determined by DKA.Â Â For example, some might want a strong particle not easily damaged during transport; others might require a lightweight powder that dissolves easily.Â Use of DKA can determine if a given feed will dry into hollow spheres, shriveled particles, ?doughnut? shapes, etc. while providing insight into design of drying equipment and process parameters for attaining a certain type of particle.

(758g) An Integrated Management System to Optimize Process Operations Within the Pharmaceutical Industry

Drayton, R. A., THS, Inc
Gloor, G. J., Imperial College London
Grills, J.,
Originally presented on: 11/7/2013 17:15:00 - 17:35:00

One of the major challenges in developing a pharmaceutical process is the need to supply material of a quality representative to the final product, and at the same time, develop a safe and scalable process that can evolve into the final manufacturing process. In order to overcome this challenge, requires careful integration of process chemistry development, analytical techniques and process engineering of the chemical synthetic route. Historically, process safety has been regarded as a manufacturing concern, leaving the collection and analysis of process safety information, such as thermal stability and thermo-chemistry to the latter stages of process development. Since, other constraints such as product impurity was developed earlier in the process, this introduced the need for engineering solutions to process safety problems later in the process. This could have been avoided by straightforward changes in process chemistry, had these problems been identified and determined earlier in development. In addition, the process safety information that is obtained from physical experiment to address the process safety problems, can also provide valuable insight into, *inter alia*, process robustness and optimum operating conditions. Thus, the opportunity to leverage this information is lost by separating process safety from process chemistry and regarding it as an independent compliance function.

It is proposed here, to integrate the activities generally associated with process safety, such as thermal stability testing, reaction engineering and safety management into the development process for pharmaceuticals and fine chemicals. This will enable the data generated by the process safety function to be part of the dossier of process information that can be utilized by chemists and engineers in developing the "operating envelope". In addition, companies can avoid the delay and expense due to process changes that result from safety recommendations at latter stages in the process, and which conflicts with the process declarations in regulatory filings. This paper presents a novel integrated management system that can optimize process operations within the pharmaceutical industry.

(759a) Simulation Study Comparing Oil Emulsion Particle Size Distributions On a Performance Basis

Stamatis, S. D., Purdue University
Kirsch, L. E., The University of Iowa
Originally presented on: 11/7/2013 15:15:00 - 15:40:00

Assessments of particle size distribution can be critical in the design and development of multiphasic liquid pharmaceutical formulations and in establishment of the pharmaceutical equivalence of generic emulsion and suspension drug products.

The assessment of particle size distribution similarity is often reduced to comparing mean particle size or span. While these numbers are easy to compare, they fail to capture any skew in the particle size distribution and the physical interpretation of these quantities is limited for multimodal distributions. Particle size distribution similarity needs be quantitatively defined in a way that rationally accounts for these limitations and further investigate if a similarity metric can be used a surrogate for in vitro release performance differences.

In this work, batch particle size distributions were defined with a three parameter log skew normal distribution. Monte Carlo sampling was used to compute the population particle size distribution to account for uncertainty in the three parameters. Systematic differences in test populations were considered to produce several test populations. Each test population's similarity to the reference population was assessed with two probabilistic metrics of distribution similarity: the overlapping coefficient (OVL) and a related measure that considers the relative location of the probability mass called PROB. Both of these metrics consider any skew in the particle size distribution as well as provide consistent meaning when either the test or reference population is multimodal. We show that the PROB measure of similarity correlates with differences in time to 25, 50 and 75% release as computed by a Higuchi type model under sink conditions and diffusion limited transport. Significance of the differences was assessed with a Wilcoxon rank sum test at 95% confidence.

(759b) Impact of Hpmcas Polymer Properties On the in Vitro Performance of Spray Dried Dispersions - A Quality By Design Approach

Porter, W. III, The Dow Chemical Company
Schmitt, R.,
Petermann, O., The Dow Chemical Company
Guillaudeu, S., The Dow Chemical Company
Brackhagen, M., The Dow Chemical Company
Sprehe, M.,
Originally presented on: 11/7/2013 15:40:00 - 16:05:00

An increasing number of new drug entities suffer from low solubility making the development of a viable deliverable dose challenging. Hypromellose acetate succinate (HPMCAS) has proven effective at producing stable amorphous solid dispersions for poorly soluble active pharmaceutical compounds that demonstrate enhanced free drug levels in solution and inhibit recrystallization. Currently, there is a limited variety of HPMCAS product grades available, covering only a subspace of the compendial range. The limited coverage of the design space impedes the ability to understand how the HPMCAS properties influence the stability and performance of an amorphous dispersion.

The present study examines the in vitro drug release rate and crystallization inhibition of amorphous spray dried dispersions formed with poorly soluble model APIs and HPMCAS polymers. HPMCAS polymers were varied in a Quality by Design approach to utilize the compendial range of acetate and succinate substitution levels and a range of molecular weights.

The effect of drug release rate and crystallization inhibition as a function of the HPMCAS acetate and succinate substitution levels in combination with the effect of the polymer molecular weight are evaluated for poorly soluble model compounds with varied physical properties. Trends linking the physical properties of the model compounds with supersaturation and crystallization inhibition based on acetate and succinate levels and the molecular weight of the HPMCAS are evaluated.

(759c) Principal Component Analysis for Pharmaceutical Process and Product Characterization:Case Study From Immediate and Controlled Release Tablet Manufacturing

Wu, H., FDA

Originally presented on: 11/7/2013 16:05:00 - 16:30:00

As an important data analysis technique for reducing dimensionality of complex process data, Principal Component Analysis (PCA) has been widely used in pharmaceutical PAT applications. The objectives of this work were to utilize PCA for: (1) differentiating physical and chemical characteristics of pharmaceutical solid dosage forms; and (2) establishing link among formulation parameters, manufacturing process variables, and pharmaceutical product quality attributes of controlled release dosage form. Immediate-release and controlled-release theophylline tablets were manufactured and subjected to NIR diffuse reflectance spectroscopy analysis. The raw or preprocessed NIR spectral datasets grouped with either constant tablet hardness, or constant formulation chemistry, or constant core tablet API strength were subjected to PCA. It was determined that the first and second principal components track tablet hardness and chemical composition for immediate release tablets, respectively. For the film-coated controlled release tablets, process trajectory for the film-coating process demonstrated an information-critical coating thickness. The controlled-release tablets were subjected to dissolution testing. A link between coating film thickness and drug dissolution rate was identified via integration of PCA and mass-transfer kinetics analysis of coated tablet dissolution process. Therefore, the possibility of using PCA to separate the physical and chemical information embedded with the NIR spectra of tablets, and to depict link among pharmaceutical formulation parameters, processing variables, and performance characteristics of tablets was demonstrated.

(759d) Dissolution of Different Commercial Aspirin Tablets Using a Novel Off-Center Paddle Impeller (OPI) Dissolution Testing System

Qu, Y.,
Armenante, P. M., New Jersey Institute of Technology
Originally presented on: 11/7/2013 16:30:00 - 16:55:00

Dissolution testing is routinely conducted in the pharmaceutical industry to provide *in vitro* drug release information for quality control purposes. The most common dissolution testing system for solid dosage forms is the United States Pharmacopeia (USP) Dissolution Testing Apparatus 2. In this work, a modified Apparatus 2, termed "OPI" System for "off-center paddle impeller," in which the impeller was placed 8 mm off center in the vessel was tested to determine its sensitivity to differentiate between the dissolution profiles of differently formulated and manufactured tablets. Dissolution tests were conducted with both the OPI System and the Standard System using three different brands of aspirin at nine different tablet positions. The OPI system produced dissolution profiles that were highly dependent on the different brands of aspirin used, similarly to those generated in the Standard System. However, the dissolution profiles obtained with the OPI apparatus were found to be largely independent of the tablet location at the vessel bottom, whereas those obtained in the Standard System generated statistically different profiles depending on tablet location. It can be concluded that the newly proposed OPI system can effectively eliminate artifacts generated by random settling of the tablet at the vessel bottom, thus making the test more robust, while at the same time being just as sensitive as the Standard System to actual differences in differently manufactured tablets having intrinsically different dissolution profiles.

(759e) The Effect of Blend Shearing and Compression On Dissolution Dynamics for Immediate Release Tablets

Callegari, G., Rutgers University
Awad, T., Rutgers University
Wang, Y., Rutgers University
Pawar, P., Rutgers University
Dendamrongvit, W., Rutgers University
Drazer, G., Rutgers University
Muzzio, F. J., Rutgers, The State University of New Jersey
Cuitino, A., Rutgers University
Originally presented on: 11/7/2013 16:55:00 - 17:20:00

Knowledge of the effect of processing parameters and formulation variations in drug release kinetics is key for assessing the risks involved in manufacturing tablets. This is the core of Quality by Design principle, the vehicle for how drugs are developed and commercially manufactured nowadays.

We study tablets compacted from blends of Micro-Cellulose and Lactose at different ratios, lubricant agent Magnesium Stearate and Acetaminophen as active ingredient. Blends are pre-processed at diverse shearing conditions and then compacted at different pressures. We investigate the effect of formulation, shearing of the blend and compaction force variations on the dissolution profiles obtained experimentally using the USP-II apparatus. We discuss the results in view of tablet characterization experiments (including SEM/EDS, X Ray Micro Computerized Tomography, Mercury Porosimetry) as well as independent dissolution experiments in which we monitor liquid penetration dynamics and tablet deformation.

(759f) Design Space Approach to Continuous Manufacturing of Drug Loaded Terapeutic Films

Huynh, P., Rutgers University
Takhistov, P., Rutgers, The State University of New Jersey
Originally presented on: 11/7/2013 17:20:00 - 17:45:00

Edible polymeric films are promising technology for the systemic delivery of active pharmaceutical ingredients (API). These films are designed as composite materials with the precise systemic dosing in safe and effective format. Development of effective delivery system for API requires a control of the structure and morphology of polymer films at the interface with API. Manufacturing of thin films is a multi-step process that controls the thickness, drug concentration, residual volatiles,
mechanical strength, and disintegration rate of therapeutic films. We present a comprehensive analysis of the factors that impacts manufacturability and performance of the oral films, such as predictive understanding of maintaining particle stability and content uniformity in multi-component film precursors at various process conditions, based on the molecular and interfacial interactions between slurry components and understanding of what properties of the individual material and their combinations are critical to film formation performance for a given process (material properties and process suitability/selection).

All maor components of the formulation, processing conditions including colloidal cstability of BSC class II drug particles are discussed. We sucsessfully demonstrate optimozation of the mixing, drying, and particle astbilization methods to optimioza film performance.

(762a) Process Development and Manufacturing: Single Use Versus Glass and Stainless Steel

Ribault, S., Merck BioDevelopment

Originally presented on: 11/7/2013 15:15:00 - 15:40:00

Single-use equipments are now commonly used for process development and production activities. The advantages of these equipments have been well demonstrated over more than a decade and include absence of cleaning, no need for sterilization, easy and fast validation, reduced labor time and increased flexibility. However, only a few data on their scalability are published and most studies focus on a single type of equipment such as Single-Use Bioreactors (SUBs) or on DSP units such as clarification or chromatography skids. Few studies actually describe full USP and DSP disposable trains.

During the period 2010-2012, we performed studies in order to evaluate the performances of Single-Use equipments from early Process Development activities down to the manufactured Drug Substance after a full manufacturing process. As different formats were available, this study compared performances of several types of equipments from 3L to 1250L scale. The evaluation was performed both for Process Development applications and for clinical material production through the development of a fed-batch process and its use for clinical scale production.

The quality of the molecule together with the titer and the cell growth were measured. The process performances were compared in single use, glass and stainless steel bioreactors of different sizes including 3L, 50L, 200L and 1,250L. Several clinical runs at 200L and 1250L scale were performed to ensure a meaningful comparison. We evaluated the differences between the systems using different types of cells such as CHO and NS0 but also different molecules including several MAbs and Fc-Fusion protein. The data generated allowed to conclude on the scalability and the comparability of the different systems and technologies. Beyond the process performances and technical data we recorded timings and evaluated the cost of the different operations to complete the assessment.

This study demonstrated the benefits of using disposable equipment in several key areas. The comparison of stainless steel, glass and disposable equipment showed how comparable they are regarding titers, molecule quality…and how different they are regarding organization and financial aspects. The operator feedbacks on the ease of use and the necessary changes in the GMP area to introduce SUBs were discussed as well.

(762b) Development and Implementation of a Single-Use, Closed Depth Filtration System for Primary Clarification

McCann, D., Biogen Idec
Originally presented on: 11/7/2013 15:40:00 - 16:05:00

Biogen Idec has developed and implemented a single-use and completely disposable clarification platform for use within its controlled non-classified flexible volume manufacturing facility in RTP, NC. This presentation demonstrates the capability of Millistak + POD single-use technologies for high cell density (TCD <25E6/mL) clarification while creating and maintaining a closed system for microbial control. Major challenges with this primary clarification operation were the large volume of WFI required (>22,000L) for priming, flushing and caustic neutralization, as well as the high flow rates of >100L/min recommended for reducing leachable/extractable (L/E) substance levels. These challenges were mitigated by assessing lower flux rates and volumes to adequately meet L/E target specs provided by the manufacturer. Furthermore, the caustic neutralization after pre-sanitization utilized capture column equilibration instead of WFI resulting in a large reduction of volume necessary for neutralization. Overall, the WFI volume was reduced by ~20,000L and facility fit constraints around flow rates was resolved with lower flux rates.

(762c) Single Use Centrifuge- CHO Clarification

Richardson, D. R., PSangelus

Originally presented on: 11/7/2013 16:05:00 - 16:30:00

Abstract Text:
The large scale economic collection of proteins is an important problem for the Biotechnology industry.
Generally, proteins are produced by an engineered cell culture where a recombinant plasmid containing the gene for that protein is inserted. Since the cell lines used are living organisms, they must be fed with a complex growth medium. Separation of the desired protein from the fermentation broth poses challenges.
Advances in cell culture techniques have greatly increased the cell populations and titers of the target proteins in culture fluid. This increase in upstream efficiency has led to a bottleneck in downstream processing at the cell-harvest stage. Cell harvesting or clarification of the cell culture fluid is an important process. When the product is extracellular, clarification is used to separate the product from the cells and cellular debris. When the finished product is the cells or intercellular proteins/virus cell harvesting is performed. Most Biopharm companies today are seeking ways to integrate this single use technology (clarification/harvest) into the single use process path.
The UniFuge single use centrifuge is fits very well into the single use initiative.
This presentation will compare the UniFuge to several cell harvest/clarification processes

(762d) Predicting Buffer and Media Preparation Times for Single-Use Mixer Scale-Up

Trafton, B., EMD Millipore
Sokolnicki, A., EMD Millipore Corp.
Felo, M., EMD Millipore Corp.
Originally presented on: 11/7/2013 16:30:00 - 16:55:00

As the implementation of single-use mixing technologies in biopharmaceutical manufacturing operations expands, scaling and mix time predictability has been a focus for characterization and improvement. For a series of commercially available single-use mixers, operating conditions consisting of solution volume, mixer power, and mixer speed were identified using a mixer scaling and predictability parameter, which produced a similar mixing time for a given solution across different scales of mixer. As exhibited in this study, this methodology was used to develop a protocol for the time predictable preparation of several buffers and cell culture media at volumes ranging from 10 to 1000 L. The development of this protocol will facilitate scale-up and scale-down during process development and process simulation efforts while reducing resource consumption. Moreover, the 50 L system demonstrated to be a reliable scale-down device for the development of time predictive mixing operations that could be scaled and implemented in the 1000 L system.

(762e) Characterization and Scalability of the Impulse Single-Use Mixing System

Pavlik, R., ASI

Originally presented on: 11/7/2013 16:55:00 - 17:20:00

Single-Use mixing systems are increasingly accepted in the production of bio-pharmaceutical therapies because of their flexibility for multiproduct facilities, speed to market and reduced risk of cross-contamination. The imPULSE Single-Use Mixing Platform from ASI combines the benefits of single-use bags with an innovative patented mixing technology. The agitator or mixing head design is unique in comparison to traditional impellers with its hexagonal shaped disc and film flaps that open on the upstroke and close on the down stroke. This achieves a vertical top to bottom mixing pattern without unwanted solid body rotation typical of traditional mixers. This simplifies design by eliminating the need for baffles or off-set agitators to maximize power tranmission from the agitator to the liquid phase. The scalable geometric principles of the reciprocating mixing disc and vessel will be presented together with characterization using a combination of mixing studies and Computational Fluid Dynamic (CFD) methodologies. The results of these studies on the imPULSE systems throughout the entire platform of bench top to 5,000 Liter vessel sizes will be discussed.

(762f) Advantages and Limitations of Single-Use Systems in Downstream Processing

Siletti, C., INTELLIGEN, INC.
Petrides, D., INTELLIGEN, INC.
Originally presented on: 11/7/2013 17:20:00 - 17:45:00

Biopharmaceutical downstream processes are generally dominated by chromatographic separations and membrane filtrations. Single-use systems can increase the efficiency and reduce the costs of these processes, but there are limitations. A common target for single use technology is the preparation, holding and distribution of the numerous buffer solutions required in downstream processing. Buffer preparation may become a capacity-limiting process (bottleneck) in certain high-capacity situations. This paper examines the use of macroscopic modeling techniques to evaluate buffer preparation for downstream processing. A series of cases studies are based on downstream processing for a 2000 L mammalian cell-culture process. Single use buffers offered a number of advantages including a 34% savings in capital costs, a 20% reduction in purified water usage, up to 20% reduction in unit and potential throughput advantages. The cost advantages decline with increased protein load on the process. Furthermore, this study explores facility floor-space limitations that become important when large numbers of disposable containers need to be managed.

AUTHOR INDEX

AUTHOR INDEX

AUTHOR INDEX

AUTHOR INDEX

AUTHOR INDEX

AUTHOR INDEX